Supply Chain Localization in the Semiconductor Industry

Tin-Chih Toly Chen

Supply Chain Localization in the Semiconductor Industry

Rebuilding the Competitiveness and Sustainability of Semiconductor Manufacturers

Springer

Tin-Chih Toly Chen
Department of Industrial Engineering
and Management
National Yang Ming Chiao Tung University
Hsinchu, Taiwan

ISBN 978-3-031-81279-8 ISBN 978-3-031-81280-4 (eBook)
https://doi.org/10.1007/978-3-031-81280-4

This Springer imprint is published by the registered company Springer Nature Switzerland AG
The registered company address is: Gewerbestrasse 11, 6330 Cham, Switzerland

Contents

Chapter 1
Semiconductor Supply Chain Management

Abstract This chapter begins by defining the concepts of supply chain and supply chain management. Eight critical processes of supply chain management are listed. Key considerations of maintaining supply chain stability and flexibility are also discussed. Subsequently, some existing supply chain management approaches are reviewed. There are four flaws in past supply chain management research: unstable collaboration among participants, delayed responses in timely communication, assumptions and constraints of simplified mathematical programming models, and lack of managerial insight. Subsequently, some existing semiconductor supply chain management methods are reviewed. Based on the review results, the problems faced by existing supply chain management methods in the semiconductor manufacturing industry are summarized. Fab localization, an unprecedented change in semiconductor manufacturing, is then highlighted. Localization in the era of globalization is different from the current trend. The former built factories close to customers at the end of the supply chain, while the latter aims to build new production capacity close to customers in the upstream of a semiconductor supply chain. Finally, several issues that need to be addressed for this type of localization are raised.

Keywords Semiconductor manufacturing · Localization · Wafer fabrication · Supply chain · Supply chain management

1.1 Supply Chains and Supply Chain Management

A **supply chain** is a network of businesses and processes that facilitate the creation, distribution and eventual sale of a product (Britt 2021). Supply chains are made up of "tiers" that depend on a supplier's proximity to the enterprise or the end product (Toikka 2023).

- Tier 1 suppliers: direct suppliers to the enterprise.
- Tier 2 suppliers: the suppliers or subcontractors of Tier 1 suppliers.
- Tier 3 suppliers: the suppliers or subcontractors of Tier 2 suppliers.

T. T. Chen, *Supply Chain Localization in the Semiconductor Industry*,
https://doi.org/10.1007/978-3-031-81280-4_1

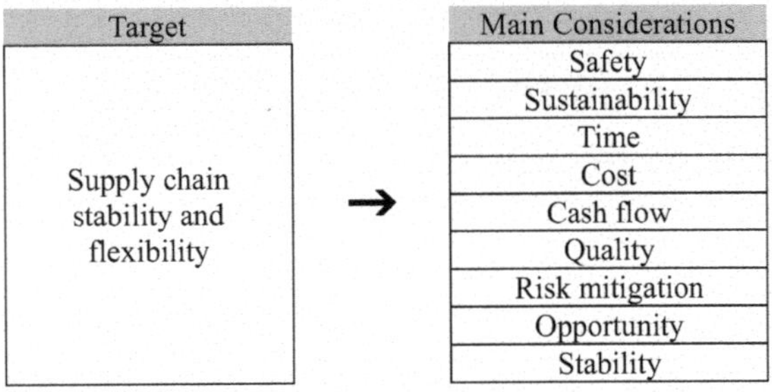

Fig. 1.1 Main considerations for maintaining supply chain stability and flexibility

The main management considerations of a supply chain include safety, sustainability, time, cost, cash flow, quality, risk mitigation, opportunity and stability (Bala 2014). In order to maintain its stability, a supply chain should also be flexible to face various emergencies (Zhang et al. 2014). Figure 1.1 shows the main considerations for maintaining supply chain stability and flexibility.

A supply can roughly be broken down into the upstream, midstream and downstream. A supply chain must optimize its upstream to downstream transportation time and costs, while also takes into account product quality. When expanding the scale of a supply chain, the risk level must still be considered. Problems encountered by any supply chain participant will affect the operations of the entire supply chain. For example, high costs incurred in a certain upstream segment will accumulate as a burden on downstream manufacturers, ultimately leading to low product profits and an unniche supply chain. On the contrary, high costs in specific downstream segments will translate into requirements for lower prices and profits of upstream suppliers. Therefore, supply chain management (SCM) is a critical task to address these issues.

Supply chain management is the design, planning, execution, control and monitoring of supply chain activities with the goals of creating net value, building a competitive infrastructure, leveraging global logistics, synchronizing supply and demand, and measuring performances globally (Dictionary 2016). Lambert (2008) mentioned that the key processes of supply chain management are.

- Customer relationship management (CRM),
- Customer service management,
- Demand management,
- Order fulfillment,
- Manufacturing flow management,
- Supplier relationship management,
- Product development and commercialization, and
- Returns management.

Most supply chain management research explores **supply chain flexibility**, **sustainability**, **risk management** and other related topics (Lam 2018).

Recently, **supply chain visibility** has gained emphasis to improve the understanding of the location/production stage and expected delivery dates of incoming products and materials so that production can be better planned (Somapa et al. 2018). In other words, companies must be able to plan orders leveraging the knowledge of potential supplies and track post-production processes through to delivery to customers. For example, chip shortages caused by the COVID-19 pandemic have force companies to change their strategies to prevent future problems. Measures taken include a combination of **supply chain mapping**, **secondary sourcing** and **digital transformation** (Chen et al. 2024a). Companies are actively stepping up their digital transformation, increase visibility with suppliers and customers, and sign more long-term agreements (Peters 2024).

1.2 Some Existing Supply Chain Management Methods

An overview of some existing supply chain management methods is as follows. First, **simulation** is one of the most common types of existing supply chain management methods (see Fig. 1.2).

AbuKhousa et al. (2014) reviewed several practices for applying **simulation** to assist healthcare supply chain management, including optimizing the distribution of scarce drugs (given constraints among clinics, pharmaceutical companies, and distribution), optimizing drugs by considering demand changes and various forms of drug inventory, and optimizing sterilization logistics in deterministic and non-deterministic scenarios. Simulation can model the operational details of a healthcare supply chain. However, many assumptions need to be made, so simulation results may differ significantly from reality. Nonetheless, simulation can be applied to show the detailed process when the production plan is executed. Simulation methods for supply chain planning and management can be further decomposed into agent-based (Chen 2024a, 2018; Kumar and Srinivasan 2010), discrete-event (Wang et al. 2023a;

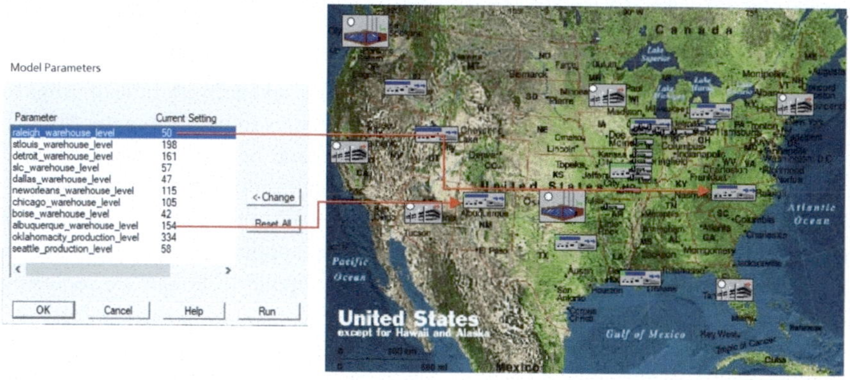

Fig. 1.2 Simulation as one of the most common types of existing supply chain management methods

Chen 2023a), or system dynamics (Wu et al. 2024; Chen 2024b) simulation methods (Ivanov 2024).

Aldrighetti et al. (2019) believed that the difference between healthcare supply chains and other supply chains lies in the high prices of materials and products and the serious consequences of the products (related to human lives). Nevertheless, like other types of supply chains, healthcare supply chains face stiff competition. They argued that eliminating waste, reducing costs, improving quality and service consistency are the most important performance measures for healthcare supply chains.

Kwon et al. (2016) asserted that healthcare supply chain management aims to manage healthcare costs while improving the quality of care. Three strategic tools were considered beneficial in achieving this goal: understanding supply chain principles, improving processes, and deploying logistics tools.

Secondly, traditional statistical techniques are widely used in supply chain management because most people have learned them. For example, Polater and Demirdogen (2018) studied the impact of supply chain integration, demand forecasting, and supplier performances on patient responsiveness in public hospitals through the mediating role of supply chain resilience. They surveyed 129 managers in public hospitals and then applied **structural equation modeling (SEM)** to analyze the collected data.

The third category of supply chain management methods is mathematical programming (MP), rooted in operations research or production management. MP is commonly used for supply chain planning and is among the most popular supply chain management research topics. Several past studies formulated **MP** models to plan supply chains. However, some of these MP models were not easy to solve. Therefore, heuristics such as genetic algorithms (GAs) (Chen and Lin 2023, 2020; Altiparmak et al. 2006) needed to be proposed to help solve these problems. Goodarzian et al. (2024a) formulated a multi-objective mixed integer-nonlinear programming (MINLP) model to establish centers for a multi-level pharmaceutical supply chain to minimize the total cost and CO2 emissions of the transportation system while maximizing the shelf life of pharmaceuticals. The MINLP problem was more easily solved by converting it into a linear problem.

Similarly, Goodarzian et al. (2021) formulated another MINLP model to plan pharmaceutical supply chains to minimize the total cost and delivery time of drugs to hospitals and pharmacies. They proposed three heuristics to help solve the MINLP problem. However, some assumptions need to be made before formulating a MP model. Furthermore, criteria may depend on each other, resulting in an unfair comparison of possible plans (Emmerich and Deutz 2018; Wu et al. 2023). Techniques such as principal component analysis (PCA) can be applied to resolve dependencies between criteria (Wang et al. 2023b). PCA is a popular technique for replacing dependent variables with new independent ones.

Goodarzian et al. (2024b) formulated a multi-objective mixed integer-linear programming (MILP) model to plan a sustainable pharmaceutical supply chain, in which the CO2 emissions caused by pharmaceutical transportation were also minimized for environmental sustainability. Supply chain planning approaches aiming to minimize makespan also helps reduce the time and amount of CO2 emitted by supply chain participants.

Similarly, Ghasemi et al. (2023) also formulated a multi-objective MILP model to plan a plasma supply chain to maximize the coverage of blood donors during periods while minimizing transportation, relocation, inventory holding, and establishment costs. They applied the ε-constraint method and the multi-objective invasive weed optimization (MOIWO) algorithm to solve the MILP problem when the supply chain scale was small and large, respectively.

In a similar study, Ghasemi et al. (2022) formulated a bi-objective MILP model to plan a blood supply chain before and after a disaster. The bi-objective MILP model minimized the total cost while maximizing the satisfaction level to meet the needs of affected areas. Decision variables included whether to establish reliable distributors, unreliable distributors, and central warehouses. They also proposed a robust optimization method to help solve the bi-objective MILP problem, where the weighted sum of the average and standard deviation of performances of all practical scenarios was minimized (Emmerich and Deutz 2018).

Furthermore, fuzzy logic has shown its potential in handling uncertainty in supply chain planning. Chiu and Chen (2022) established a ubiquitous medical network with multiple 3D printing facilities to manufacture dentures. The ubiquitous medical network also worked with home delivery services, which have become increasingly popular during the COVID-19 pandemic, to save transportation time. Furthermore, interval type-II trapezoidal fuzzy numbers were used to model printing and transportation times and consider their uncertainty. An interval type-II fuzzy mixed integer-linear programming (FMILP) model was formulated and optimized to plan the operations in the ubiquitous medical network. Fuzzy logic has shown its effectiveness in modeling operational uncertainty, which is unavoidable in supply chains.

The next mainstream of supply chain management methods is MCDM. Unlike Ghasemi et al. (2023), supply chain planning approaches from the MCDM perspective can be applied to various supply chains regardless of their scales. Chen et al. (2024a) proposed an evolutionary fuzzy approach from the perspective of MCDM to improve the effectiveness of planning pharmaceutical or healthcare supply chains. Past research mostly used simulation or optimization methods, which were time-consuming or difficult to obtain the optimal solution. Key factors considered in their study included product quality, capacity, availability, experiences and relationships, and product costs. In their study, the calibrated fuzzy inverse sum of columns (cFICSM) method was designed to derive the fuzzy priorities of criteria. Fuzzy principal component analysis (FPCA) was then proposed to resolve the dependencies between criteria. Based on the new independent criteria, Fuzzy VIseKriterijumska Optimizacija I Kompromisno Resenje (FVIKOR) was applied to evaluate and compare supply chain partners. Finally, the required pieces of an order were distributed among supply chain partners. The evolutionary fuzzy approach has been applied to a pharmaceutical or healthcare supply chain for planning the production of medical masks in central Taiwan during the COVID-19 pandemic.

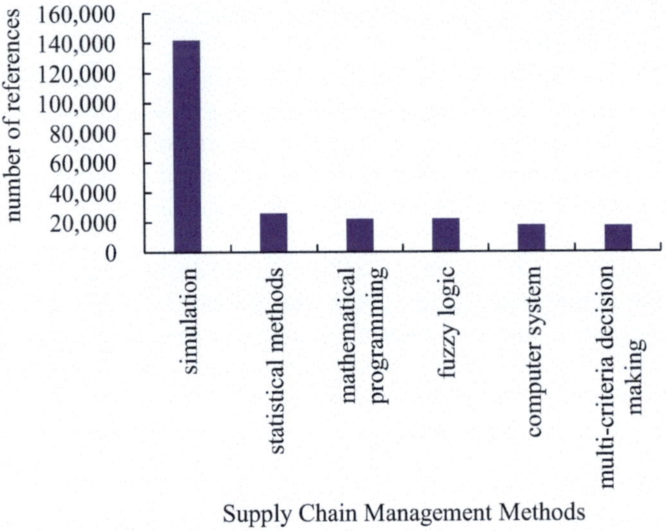

Fig. 1.3 Number of references about the applications of various types of methods in supply chain management from 2014 to 2024 (*Data source* Google Scholar)

Supply chain network optimization allows to determine optimal locations of several warehouses and factories along with the associated product flows in the supply chain (Ivanov 2024; Chen 2024c, 2023b; Matinrad et al. 2013; Eskandarpour et al. 2015).

A systematic approach to managing supply chain operations (i.e., execution, negotiation and control) is clearly the most effective and efficient. For example, SEMI launched the Supply Chain Initiative to bring together and share best practices aimed at increasing supply chain agility and resilience (Peters 2024).

Figure 1.3 provides statistics on the popularity of various types of methods in supply chain management. Clearly, simulation, statistical methods, and mathematical programming are the most popular supply chain management methods.

1.2.1 Problems with Existing Methods

Most previous research on supply chain management was focused on the modeling, simulation, design, and evaluation of supply chains. Digitization (such as acquiring an enterprise resource planning (ERP) or other information systems) is the most common way to plan supply chains in many industries (Lambert 2008). Existing supply chain planning methods have the following research gaps to be filled (see Fig. 1.4):

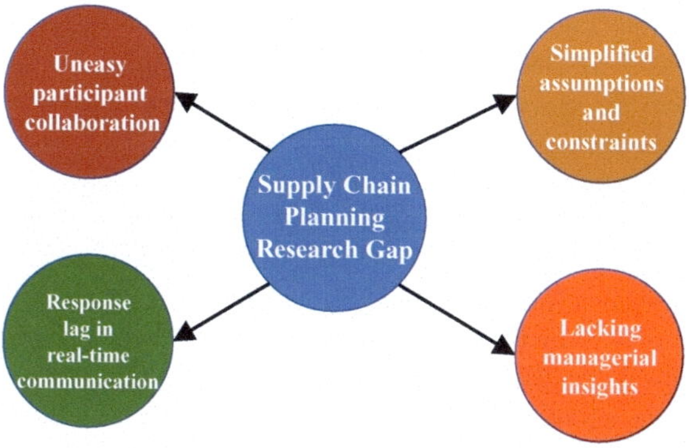

Fig. 1.4 Research gap of existing supply chain planning methods

- The effectiveness of a supply chain depends on the collaboration of participants, a management issue that cannot be easily addressed.
- Real-time communication methods assume that supply chain partners will take immediate actions when needed, which is unrealistic because cross-supplier collaboration takes much time to ensure that these actions are acceptable.
- Optimization-driven methods identify a feasible way for supply chain participants to collaborate on achieving quantifiable goals. However, such methods are based on simplified assumptions and constraints, so the optimization result can only be used as a reference. Furthermore, the optimized plan may contain fragmented actions that supply chain participants cannot take.
- Managerial methods, although less discussed and often unoptimized, are more practical.

1.3 Semiconductor Supply Chains

Semiconductors have become indispensable electronic components for daily life (Chen and Lee 2024), so there are many related supply chains (Mönch et al. 2018a). With the advancement of semiconductor, information, networking and communication technologies, the upstream, midstream and downstream of a semiconductor supply chain can be spread all over the world, which also affects the development of other industries (Chen 2023c). The upstream of semiconductor supply chains covers silicon intellectual property (IP) design and integrated circuit (IC) design industries; the midstream includes semiconductor raw materials, semiconductor equipment, photomasks, chemicals, IC manufacturing and other industries; the downstream includes industries such as assembly, packaging and testing (APT) and IC sales channels (Jiang 2024). Figure 1.5 shows the parts of a semiconductor supply chain (Taipei Exchange, Introduction to the semiconductor industry chain 2024; Chen and Wang 2021). According to TSMC, the world's largest foundry, a semiconductor supply chain starts from system companies, passes through electronic manufacturing services (EMSs), IC design, IC manufacturing, IC packaging and testing, end consumers, and finally returns to system companies (TSMC 2024) (see Fig. 1.6). EMS includes companies that design, manufacture, test, distribute, and provide return/repair services of electronic components and assemblies for original equipment manufacturers (OEMs) (Wu 2013).

Upstream IP design and IC design are the functions of designing chips, including logic design, circuit design and wafer design. They are also the frontend of IC manufacturing. Well-known IC design companies include Nvidia, Advanced Micro Devices (AMD), MediaTek, Novatek, etc. After IC design, the midstream of the supply chain is IC manufacturing, for which the first step is wafer fabrication (University Wafer 2024), which requires raw materials, equipment, masks and related chemicals before processing. Raw materials mainly include silicon wafers, thin film deposition materials and photolithography materials (University Wafer 2024). Using these raw materials, basic circuit patterns are printed on wafers according to product design, and then oxidized, diffused, chemical vapor deposited, etched, ion implanted, etc., to

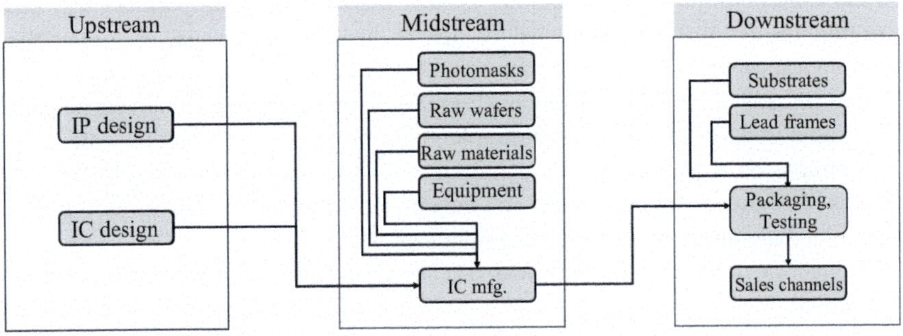

Fig. 1.5 Parts of a semiconductor supply chain

Fig. 1.6 TSMC's definition of semiconductor supply chains

fabricate circuits and components. Top companies (foundries) in wafer fabrication include Taiwan Semiconductor Manufacturing Co. (TSMC), Samsung and GlobalFoundries (Electronics 2023). The most downstream part of a semiconductor supply chain is assembly, IC packaging and testing (Deng et al. 2010). IC testing requires substrates and lead frames. IC packaging involves cutting the processed wafers and then packaging them with materials such as plastic, ceramic or metal to protect the chips. IC testing is to test the electrical properties of processed wafers, confirm the normal functions of chips and check yield to ensure product quality (Chen and Chiu 2018; Wang and Chen 2023). Top IC packaging and testing companies include Advanced Semiconductor Engineering, Amkor Technology, etc. (Smart Electronics Industry Project Promotion Office 2017).

1.3.1 Taiwan Semiconductor Supply Chains

Taiwan has a long development history in semiconductor manufacturing. Taiwan's semiconductor industry provided downstream semiconductor packaging and testing services in 1974 and officially became a part of U.S. semiconductor supply chains. In 1979, Hsinchu Science Park was established, and Industrial Technology Research Institute (ITRI) introduced IC design and manufacturing technologies to enable wafer R&D and fabrication to develop vigorously (The Economist, Taiwan's dominance of the chip industry makes it more important 2023).

Due to the long development history and clustering effect in Taiwan, the quantity and quality of supplied wafers lead the world. Taiwan has become an indispensable role in semiconductor supply chains. In 1980, United Microelectronics Corporation (UMC) was established as the first wafer foundry in Taiwan. The company's target market is chips with high demand, while advanced processes are not the primary development focus. TSMC, established in 1987, has ranked first in the world in terms of market share and is world-famous for wafer foundry together with other companies. In addition to TSMC and UMC, there are many other IC manufacturing companies in Taiwan, such as Vanguard International Semiconductor Corporation, Ltd., Powerchip Semiconductor Manufacturing Corporation, Ltd., etc. Most of them have factories in science parks, which shows the clustering effect on the development advantages of the semiconductor industry in Taiwan.

The total export value of the semiconductor industry in Taiwan has grown year by year, making it one of the most important industries. According to the statistics from the Accounting Office of the Ministry of Finance of Taiwan, the total export value of IC was US$78.2 billion in 2016 and US$155.5 billion in 2021. The amount has more than doubled in five years (Wu 2014). According to the statistics from the IEK Consulting of Industrial Technology Research Institute, Taiwan's semiconductor industry has a global market share of 20.1%, and its output value ranks first in the world (Wu 2014). Figure 1.7 shows the total value and annual growth rate of Taiwan's IC exports in recent years (The Directorate General of Budget, Accounting and Statistics 2011).

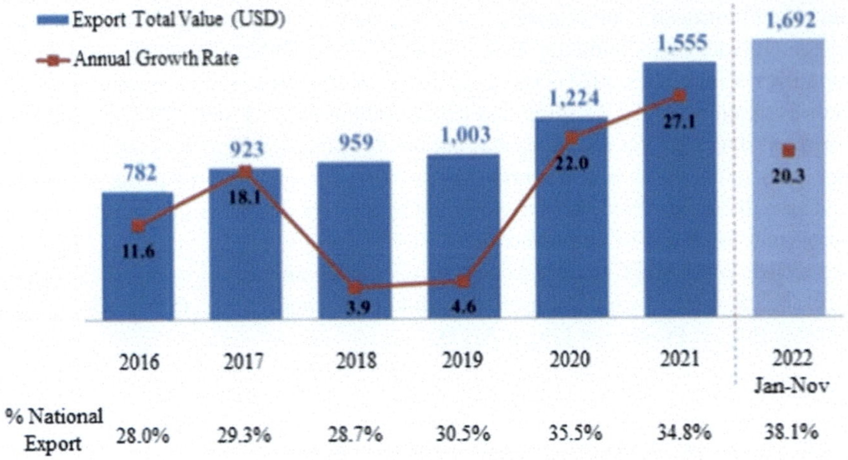

Fig. 1.7 Total value and annual growth rate of Taiwan's IC exports in recent years

1.4 Semiconductor Supply Chain Management

According to TSMC, the world's largest foundry, the four principles of supply chain management are code compliance, risk assessment, participation in audits, and continuous improvement. To make semiconductor supply chains sustainable, companies need to pay attention to environmental processes, social practices, and economics (Jiang 2024).

Wang et al. (2007) compared the differences between supply chain management in the semiconductor manufacturing industry and those in other industries. The comparison showed that supply chain management in semiconductor manufacturing is more challenging because of the highly stochastic and nonlinear nature of various environmental and production conditions such as cycle time, unit cost, yield, and customer demand. Therefore, much past research on semiconductor manufacturing management has focused on resolving this uncertainty in order to better consider the impact of such conditions (Chen et al. 2024b; Wang et al. 2023c; Chen and Chiu 2021; Lin and Chen 2019a).

Kempf (2004) proposed a control-oriented approach for supply chain management in semiconductor manufacturing, which was formulated as a nonlinear stochastic combinatorial optimization problem. In the control-oriented approach, the outer loop sought to address the planning part of the problem, while the inner loop attempted to manage the execution aspect. In addition, a linear programming (LP) model was optimized to generate the material release plan and a dynamic programming (DP) model was optimized to set inventory safety stock targets in the outer loop. The LP model maximized the net profit over the planning period, calculated as sales minus fabrication costs, inventory holding costs (Chen 2023d) and penalties for overstock and deviation from inventory targets. Constraints were the balance between fab inputs, inventory and output, and the comparison between delivered quantities and customer demand.

Aelker et al. (2013) argued that semiconductor supply chains are predictable in theory, but in practice, the nonlinear effects of many causal relationships make the system difficult to predict. Semiconductor supply chains are also extremely sensitive to initial conditions, so very small changes in one condition can lead to completely different consequences. Therefore, semiconductor supply chains were classified as emerging, self-organizing, dynamic and evolving complex adaptive systems (CAS).

Mönch et al. (2018b) highlighted the differences in values in the semiconductor supply chain. The upstream operations of semiconductor manufacturing (such as wafer fabrication and probe) are conducted in highly industrialized countries, while the downstream operations (assembly, packaging and testing) are conducted in countries where labor is cheap. Such differences in values make the cooperation among semiconductor supply chain participants difficult. Other difficulties in managing semiconductor supply chains include demand uncertainty, product line diversification across various IC applications, and shortened product life cycles.

Wang and Rivera (2008) proposed a model predictive control (MPC) algorithm to plan a semiconductor supply chain, which minimized the sum of daily inventory levels, daily production levels, and their changes. The planning horizon was quite short-term. Control-based supply chain approaches seem less practical for medium- or long-term planning because they expect different supply chain participants to perform these actions according to the control rules immediately, which is unlikely in practice. Delays due to human intervention in cross-organizational operations are not uncommon.

Lee et al. (2006) formulated several MP models to plan the production and distribution functions of a semiconductor supply chain. These MP models were arranged in a hierarchical manner and solved sequentially. The optimal solution of MP models in upper layers became the constraints of MP models in lower layers. In addition, the MP models for the production function were optimized before those for the distribution function.

Chien et al. (2024) reviewed the relevant literature on semiconductor supply chain management and proposed a new planning framework that addresses several issues such as price strategy, demand planning, capacity mix, capital expenditure and cost structure to improve overall returns. They believed that following such a planning framework was conducive to building a sustainable semiconductor supply chain.

Chen et al. (2016) formulated a MINLP model (because of the use of the max() function) for planning the production and logistics operations of a supply chain of integrated device manufacturers (IDMs). The objective function maximized supply chain profits within the planning horizon by considering sales prices, product quality, delivery quantity, shortage (penalty) cost, production cost, transportation cost, order processing costs and inventory holding costs. The semiconductor supply chain was modeled as a push–pull system and a decoupling point was identified. The production plan before the decoupling point needed to meet the forecasted inventory level at the decoupling point. The logistics plan needed to satisfy customer orders after the decoupling point. The concept of a decoupling point is similar to that of a pacemaker in lean manufacturing (Chen and Wang 2023). In fact, lean manufacturing has been extensively applied in the semiconductor manufacturing industry (Doolen and Hacker 2005; Chen and Wang 2022; Goss et al. 2010; Chen 2023e).

The existing methods for supply chain management in the semiconductor manufacturing industry has the following problems:

- Most existing methods formulated MP models. However, many details are omitted and many assumptions need to be made to simplify the problem. In addition, some MP models are NP-hard and difficult to obtain global optimal solutions.
- In contrast, simulation-based methods can preserve many details. However, even so, the simulated semiconductor supply chain still differs significantly from the real one. This is partly due to the reluctance of supply chain players to disclose the details of their operations.
- Instead of using real case data, a numerical example was often generated to illustrate the applicability of the proposed methodology. This difficulty is not uncommon because real semiconductor supply chain planning problems often involve large amounts of confidential information.

Figure 1.8 summarizes the difficulties of semiconductor supply chain management.

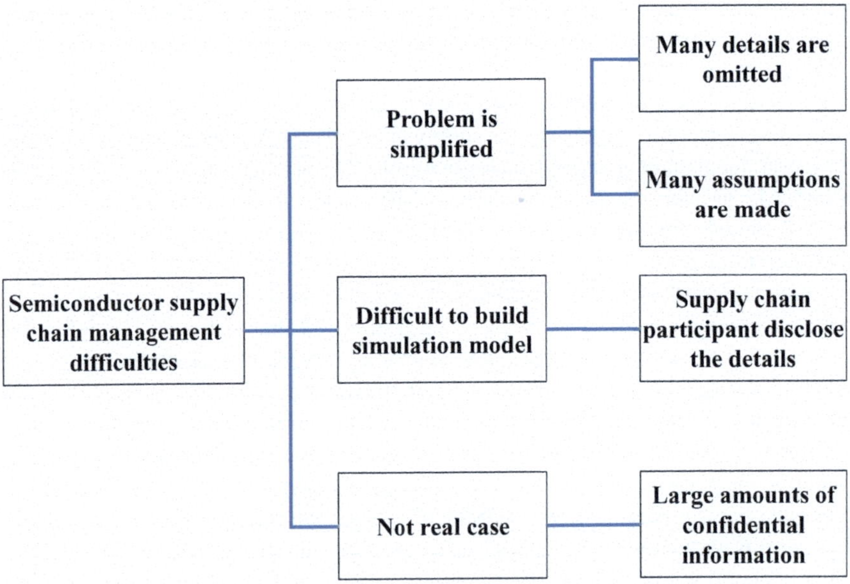

Fig. 1.8 Difficulties of semiconductor supply chain management

1.5 Semiconductor Supply Chain Localization

Recently, an unprecedented change in semiconductor manufacturing has been the **localization** of wafer fabs (Chen et al. 2023). For example, TSMC has begun building a new wafer fab in Phoenix, Arizona, USA since 2020 (Shilov 2022). Around the same time, TSMC also began building a new wafer fab in Kumamoto, Japan (Chiang 2024), whereas in the past most of its wafer fabs were located in Taiwan. TSMC's main rival Samsung has also begun construction of a new wafer fab in Arizona, USA (Shilov 2023). This behavior is called "localization" of wafer fabrication capacity (or semiconductor supply chains) and has distinctive characteristics that originate from the requirements of customers (i.e., chip designers). The main reasons for the localization of a semiconductor supply chain include.

- US-China trade war,
- Geopolitics,
- The COVID-19 pandemic,
- Ukrainian-Russian war,
- The rise of environmental awareness, and
- The need for semiconductor supply chain transparency.

In the past, localization usually meant establishing factories close to customers at the end of a supply chain in the era of globalization (Wu and Jia 2018). The recent move to localize a semiconductor supply chain is also aimed at building new production capacity close to customers but upstream in the semiconductor supply chain. These new localized wafer fabs can be expected to have higher manufacturing costs and slower learning rates. However, new localized wafer fabs can also avoid the impact of possible wars, relieve pressure from local governments, and create new cooperation opportunities.

As mentioned earlier, changes to any part of a supply chain can have a significant impact on the supply chain. So far, backend packaging and testing companies have also decided to build new capacity near localized wafer fabs (TechNews 2020). Additional impacts on the semiconductor supply chain need to be observed.

The following issues will arise in this type of localization to be discussed (Fig. 1.9):

- What impact will these localization actions have on the foundries' competitiveness (Chen 2023f)?
- Will the sustainability of these foundries be enhanced as a result?
- How will the competitors of these foundries respond?
- Will such localization actions affect the competitiveness or sustainability of the related semiconductor supply chains?
- Will the scale of such localization actions expand, for example, into markets with mature wafer fabrication processes?

The localization of semiconductor supply chains discussed in this book is different from the localization actions taken by the Chinese semiconductor manufacturing

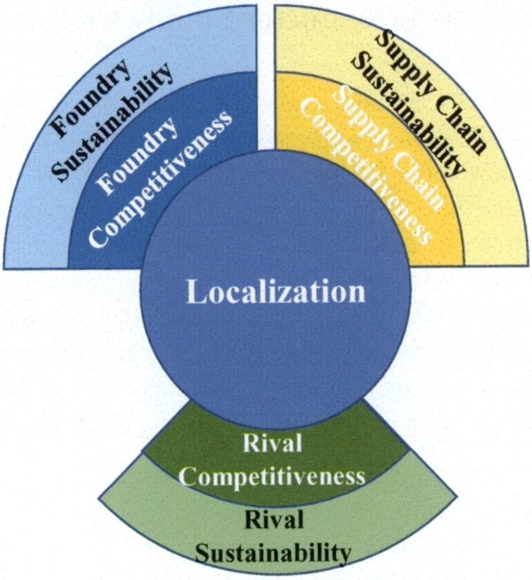

Fig. 1.9 Issues with semiconductor supply chain localization

industry. Major Chinese chipmakers such as Semiconductor Manufacturing International Corporation (SMIC) and Changxin Memory Technology (CXMT) are working to localize the supply of key chip materials and chemicals (TrendForce 2024). The move is expected to offset U.S. export controls and could exclude global suppliers from the Chinese market. The localization efforts taken by Chinese semiconductor manufacturing industry initially applied to less advanced wafer fabrication processes, such as 55 nm and 40 nm, but will eventually be extended to processes below 28 nm. However, although these localized actions have limited impact on semiconductor manufacturers in other countries, they have somewhat helped improve the competitiveness and sustainable development of local semiconductor manufacturers in China. Most of the fields that Chinese semiconductor manufacturers can conquer are low-end products, while mid- to high-end products still rely heavily on foreign suppliers for the time being.

Although the current trends in the localization of semiconductor supply chains in different regions are different, the reasons behind them are the same.

1.6 Organization of This Book

Semiconductor manufacturing relies on the collaboration of many companies distributed in the upstream, middle, and downstream of a semiconductor supply chain. Under the impact of globalization, the participants of a semiconductor supply chain are located all over the world, forming a "long chain" according to the unequal advantages of different regions in order to reduce production costs and improve operational efficiency. In recent years, supply chains in several industries adopt long chains and short chains in parallel. The semiconductor industry is no exception. In addition to globalization, they also tend to localize production, which means moving capacity back to the home country of chip designers.

This book intends to discuss the localization of semiconductor supply chains, including how manufacturers and their customers respond, how to formulate localization strategies, and how to rebuild the competitiveness and sustainability of semiconductor manufacturers through supply chain localization.

In specific, the outline of the present book is structured as follows.

This chapter first defines the concepts of supply chain and supply chain management. Eight key processes of supply chain management are listed. The main considerations of maintaining supply chain stability and flexibility are also discussed. Subsequently, some existing supply chain management approaches are reviewed. Past supply chain management research suffers from four flaws: unstable collaboration among participants, delayed responses in just-in-time communication, simplifying assumptions and constraints in MP models, and lack of managerial insight. Taiwan's semiconductor manufacturing industry was taken as an example to describe the structure of a semiconductor supply chain. Subsequently, some existing semiconductor supply chain management methods are reviewed. Based on the review results, the problems faced by existing supply chain management methods in the semiconductor manufacturing industry are summarized. The final subsection focuses on the localization of wafer fabs, an unprecedented change in semiconductor manufacturing. Localization in the era of globalization is different from the current trend. The former built factories close to customers at the end of a supply chain, while the latter aims to build new production capacity close to customers in the upstream of a semiconductor supply chain. Several issues that need to be addressed in this type of localization are raised at last.

Chapter 2, Localization of Semiconductor Supply Chains—Driving Forces and Challenges, first examines the stability of a semiconductor supply chain based on two theories: the triple bottom line and the multi-generational philosophy. The sustainability of a semiconductor supply chain is clearly integral to its stability. To pursue the sustainability of a semiconductor supply chain, localization is considered a tactical strategy for wafer foundries. The current localization trend is then compared to the traditional one to show its uniqueness. Subsequently, various forces driving the localization of semiconductor supply chains are identified as the U.S.-China trade war, geopolitics, the COVID-19 pandemic, the Ukraine-Russia war, environmental awareness, and the need for semiconductor supply chain transparency. The impact of each driving force is then discussed individually. In addition, several challenges that need to be overcome in the localization of semiconductor supply chains are also listed, including different laws, cultural differences, insufficient local human resources, uncertainty about local government subsidies, competition from local competitors, and higher wage levels. The procedure for semiconductor supply chain localization is subsequently established. Finally, possible future developments in semiconductor supply chain localization are discussed, with the impact of generative artificial intelligence (GenAI) (Treppner et al. 2022; Chen et al. 2024c; Ross et al. 2021; Wang and Chen 2024) and chip-on-wafer-on-substrate (CoWoS) (WikiChip 2024) being also mentioned.

Chapter 3, Issues in Managing Localized Semiconductor Supply Chains, first emphasizes that localization is one of the important factors for the sustainable development of a supply chain. The advantages of localization mentioned in the supply chain management literature are then summarized. Several related issues in semiconductor supply chain management are also raised. The first is how to choose the most suitable factory location to localize a wafer fab. For comparison purposes, the literature on facility location selection in traditional supply chains is also reviewed. Several real cases are also analyzed. The analysis results show that the country (or region) chosen to build a localized wafer fab is actually irrelevant. Instead, the most suitable city or location in the intended country (or region) must be selected to build the localized wafer fab. The second issue is how to mitigate the impact of cultural differences. To this end, a number of measures have been proposed, including seeking the assistance from local communities, the cooperation between fabs in different time zones, the natural replacement of domestic workers, hiring more local employees, designing notification or reminder forms, and a staggered shift system. The next issue discusses the planning horizon of a semiconductor supply chain localization plan, for which the concept of flexible long-term localization is proposed. The final issue is about the wafer technology node that a localized wafer fab should adopt to meet the needs of local customers, foundries and local governments.

Chapter 4, Qualitative and Quantitative Analysis of Semiconductor Supply Chain Localization, applies some qualitative and quantitative methods to the analysis of semiconductor supply chain localization. First, SWOT analysis, one of the most well-known qualitative methods for formulating and evaluating strategies (Chen et al. 2024d), is reviewed. Some examples of SWOT analysis applications in various fields are also given. To demonstrate the application of SWOT analysis in the field of semiconductor supply chain localization, a SWOT analysis is conducted on a major foundry in the world to rationalize the foundry's past choice of building wafer fabs domestically. Subsequently, a SWOT analysis is used again to examine the foundry's localization attempts around the world. A comparative analysis of the localization plans of its two wafer fabs in two countries is also conducted. In the second part of this chapter, two MCDM methods, analytical hierarchy process (AHP) (Wu et al. 2020a, 2020b; Chen and Lin 2022; Lin and Chen 2019b; Chen 2023g) and the technique for order preference by similarity to ideal solution (TOPSIS) (Chen et al. 2023) are combined to compare the suitabilities of several locations for building the localized wafer fab. To further enhance synergies, SWOT analysis and the AHP-TOPSIS method are jointly applied to qualitatively and quantitatively assess the strengths and weaknesses of multiple locations for building the localized wafer fab.

Chapter 5, Rebuilding Semiconductor Manufacturing Competitiveness and Sustainability through Supply Chain Localization, first introduces the concept of competitiveness (Chen and Wang 2024; Chen et al. 2024e) and identifies the factors that affect a company's competitiveness from a holistic perspective. Subsequently, supply chain competitiveness is defined and the literature on factors affecting supply chain competitiveness is reviewed. The competitiveness of semiconductor manufacturing then comes into focus. Qualitative and quantitative methods for assessing the competitiveness of semiconductor manufacturers are introduced. To illustrate these assessment methods, the impact of recent fab localization initiatives on the competitiveness of these foundries is assessed. Based on the assessment results, the competitivenesses of semiconductor manufacturers in various countries are compared. The second part of this chapter introduces the concept of sustainability (Martins and Pato 2019; Chen 2023h, 2023i; Giannakis and Papadopoulos 2016). A company's short-term competitiveness is often thought to contribute to its long-term sustainability. However, some recent cases have shown that it is possible to pursue long-term sustainability (Fan et al. 2009; Chen 2023j) at the expense of short-term competitiveness (Čadil et al. 2017; Chen et al. 2020; Chen 2023k), such as the recent supply chain localization initiatives of some foundries. Nonetheless, the chapter concludes with a discussion of how to enhance the competitiveness and sustainability of semiconductor manufacturing through localization of supply chains.

References

E. AbuKhousa, J. Al-Jaroodi, S. Lazarova-Molnar, N. Mohamed, Simulation and modeling efforts to support decision making in healthcare supply chain management. Scient. World J. **2014**, 354246 (2014)

J. Aelker, T. Bauernhansl, H. Ehm, Managing complexity in supply chains: a discussion of current approaches on the example of the semiconductor industry. Proc. CIRP **7**, 79–84 (2013)

R. Aldrighetti, I. Zennaro, S. Finco, D. Battini, Healthcare supply chain simulation with disruption considerations: a case study from Northern Italy. Glob. J. Flex. Syst. Manag. **20**(Suppl 1), 81–102 (2019)

F. Altiparmak, M. Gen, L. Lin, T. Paksoy, A genetic algorithm approach for multi-objective optimization of supply chain networks. Comput. Ind. Eng. **51**(1), 196–215 (2006)

APICS Dictionary, Supply chain management (SCM) (2016). http://www.apics.org/dictionary/dictionary-information?ID=4202

K. Bala, Supply chain management: some issues and challenges-A review. Int. J. Current Eng. Technol. **4**(2), 946–953 (2014)

H. Britt, What is "upstream" and "downstream" in supply chain management? (2021). https://www.thomasnet.com/insights/what-is-upstream-downstream-supply-chain-management/

J. Čadil, K. Mirošník, J. Rehák, The lack of short-term impact of cohesion policy on the competitiveness of SMEs. Int. Small Bus. J. **35**(8), 991–1009 (2017)

T. Chen, Estimating semiconductor unit cost using an agent-based fuzzy collaborative intelligence approach with entropy-consensus. Appl. Soft Comput. **73**, 884–897 (2018)

T. Chen, M.-C. Chiu, A multi-granularity approach for estimating the sustainability of a factory simulation model—Semiconductor packaging as an example. Operat. Res.-an Int. J. **18**(3), 711–729 (2018)

T. Chen, M.-C. Chiu, An interval fuzzy number-based fuzzy collaborative forecasting approach for DRAM yield forecasting. Complex and Intell. Syst. **7**, 111–122 (2021)

T. Chen, C.-W. Lin, Smart and automation technologies for ensuring the long-term operation of a factory amid the COVID-19 pandemic: an evolving fuzzy assessment approach. Int. J. Adv. Manuf. Technol. **111**, 3545–3558 (2020)

T.-C.T. Chen, C.-W. Lin, Assessing cloud manufacturing applications using an optimally rectified FAHP approach. Complex and Intell. Syst. **8**, 5087–5099 (2022)

T. Chen, C.-W. Lin, Optimizing 3D printing facility selection for ubiquitous manufacturing using an evolving fuzzy big data analytics approach. Int. J. Adv. Manuf. Technol. **127**, 4111–4121 (2023)

T.C.T. Chen, Y.C. Wang, Fuzzy dynamic-prioritization agent-based system for forecasting job cycle time in a wafer fabrication plant. Complex and Intell. Syst. **7**(4), 2141–2154 (2021)

T.C.T. Chen, Y.C. Wang, Evaluating innovative future robotic applications in manufacturing using a fuzzy collaborative intelligence approach. The Int. J. Adv. Manufact. Technol. **130**(11), 6027–6041 (2024)

M.C. Chen, Y.H. Hsiao, H.Y. Huang, Semiconductor supply chain planning with decisions of decoupling point and VMI scenario. IEEE Trans. Syst. Man, and Cybernet.: Syst. **47**(5), 856–868 (2016)

T. Chen, Y.C. Wang, P.H. Jiang, A selectively calibrated derivation technique and generalized fuzzy TOPSIS for semiconductor supply chain localization assessment. Decision Anal. J. **8**, 100275 (2023)

T.C.T. Chen, H.C. Wu, M.C. Chiu, A deep neural network with modified random forest incremental interpretation approach for diagnosing diabetes in smart healthcare. Appl. Soft Comput. **152**, 111183 (2024b)

T. Chen, C.-W. Lin, Y.-C. Lin, A fuzzy collaborative forecasting approach based on XAI applications for cycle time range estimation. Appl. Soft Comput. **151**, 111122 (2024c)

T.C.T. Chen, H.C. Wu, K.W. Hsu, Recommending suitable hotels to travelers in the post-COVID-19 pandemic using a novel FAHP-fuzzy TOPSIS approach. Complex and Intell. Syst. **10**, 6901–6915 (2024d)

T.C.T. Chen, Job sequencing and scheduling. in *Production Planning and Control in Semiconductor Manufacturing: Big Data Analytics and Industry 4.0 Applications* (2023), pp. 77–100

T.C.T. Chen, Applications of XAI for decision making in the manufacturing domain. in *Explainable Artificial Intelligence (XAI) in Manufacturing: Methodology, Tools, and Applications* (2023), pp. 51–81

T.C.T. Chen, Big data analytics for semiconductor manufacturing. in *Production Planning and Control in Semiconductor Manufacturing: Big Data Analytics and Industry 4.0 Applications* (2023), pp. 1–19

T.C.T. Chen, Applications of XAI to job sequencing and scheduling in manufacturing. in *Explainable Artificial Intelligence (XAI) in Manufacturing: Methodology, Tools, and Applications* (2023), pp. 83–105

T.C.T. Chen, Industry 4.0 for semiconductor manufacturing. in *Production Planning and Control in Semiconductor Manufacturing: Big Data Analytics and Industry 4.0 Applications* (2023), pp. 21–40

T.C.T. Chen, Applications of XAI for forecasting in the manufacturing domain. in *Explainable Artificial Intelligence (XAI) in Manufacturing: Methodology, Tools, and Applications* (2023), pp. 1–11

T.C.T. Chen, Applications of XAI for forecasting in the manufacturing domain. in *Explainable Artificial Intelligence (XAI) in Manufacturing: Methodology, Tools, and Applications* (2023), pp. 13–50

T.C.T. Chen, Sustainable smart healthcare applications: lessons learned from the COVID-19 pandemic. in *Sustainable Smart Healthcare: Lessons Learned from the COVID-19 Pandemic* (2023), pp. 65–92

T.C.T. Chen, Enhancing the sustainability of smart healthcare applications with XAI. in *Sustainable Smart Healthcare: Lessons Learned from the COVID-19 Pandemic* (2023), pp. 93–110

T.C.T. Chen, Cycle time prediction and output projection. in *Production Planning and Control in Semiconductor Manufacturing: Big Data Analytics and Industry 4.0 Applications* (2023), pp. 41–62

T.C.T. Chen, Defect pattern analysis, yield learning modeling, and yield prediction. in *Production Planning and Control in Semiconductor Manufacturing: Big Data Analytics and Industry 4.0 Applications* (2023), pp. 63–76

T.C.T. Chen, XAmI applications to smart homes. in *Explainable Ambient Intelligence (XAmI) Explainable Artificial Intelligence Applications in Smart Life* (2024), pp. 39–61
T.C.T. Chen, Produktions-und transportplanung für ein allgegenwärtiges fertigungssystem basierend auf dreidimensionalem druck. in *3D-Druck und allgegenwärtige Fertigung* (2024), pp. 65–83
T.C.T. Chen, XAmI applications to location-aware services. in *Explainable Ambient Intelligence (XAmI) Explainable Artificial Intelligence Applications in Smart Life* (2024), pp. 63–83
T.C.T. Chen, Y.-J. Lee, Smart technologies for fall detection and prevention. in *Smart and Healthy Walking: Toward Better Health and Life in Smart Cities* (2024), pp. 73–98
T.C.T. Chen, Y.C. Wang, AI applications to shop floor management in lean manufacturing. in *Artificial Intelligence and Lean Manufacturing* (2022), pp. 75–90
T.C.T. Chen, Y.C. Wang, Grundlagen im lean management. in *Künstliche Intelligenz und schlanke Produktion* (2023), pp. 1–13
T.C.T. Chen, W.A. Chaovalitwongse, M.J. O'grady, K. Honda, Editorial: smart technologies for improving the quality of mobile health care. Health Care Managem. Sci. **23**, 171–172 (2020)
T. Chen, Y.-C. Wang, Y.-C. Lin, Production planning and control of a pharmaceutical supply chain using evolutionary fuzzy approach. J. Eng. Res. 1–10 (2024)
T.C.T. Chen, Y.C. Wang, C.W. Lin, Enhancing the effectiveness of output projection in wafer fabrication using an industry 4.0 and XAI approach. The Int. J. Adv. Manufact. Technol. **134**(1), 113–125 (2024)
S. Chiang, Chip giant TSMC opens first Japan factory as it diversifies away from Taiwan amid U.S.-China tensions (2024). https://www.cnbc.com/2024/02/26/tsmc-opens-new-plant-in-japan-as-it-diversifies-away-from-taiwan.html
C.F. Chien, H.A. Kuo, Y.S. Lin, Smart semiconductor manufacturing for pricing, demand planning, capacity portfolio and cost for sustainable supply chain management. Int J Log Res Appl **27**(1), 193–216 (2024)
M.C. Chiu, T.C.T. Chen, A ubiquitous healthcare system of 3D printing facilities for making dentures: application of type-II fuzzy logic. Digital Health **34**, 20552076221092540 (2022)
Y. Deng, J.F. Bard, G.R. Chacon, J. Stuber, Scheduling back-end operations in semiconductor manufacturing. IEEE Trans. Semicond. Manuf. **23**(2), 210–220 (2010)
T.L. Doolen, M.E. Hacker, A review of lean assessment in organizations: an exploratory study of lean practices by electronics manufacturers. J. Manuf. Syst. **24**(1), 55–67 (2005)
M.T. Emmerich, A.H. Deutz, A tutorial on multiobjective optimization: fundamentals and evolutionary methods. Nat. Comput. **17**, 585–609 (2018)
M. Eskandarpour, P. Dejax, J. Miemczyk, O. Péton, Sustainable supply chain network design: an optimization-oriented review. Omega **54**, 11–32 (2015)
M. Fan, S. Kumar, A.B. Whinston, Short-term and long-term competition between providers of shrink-wrap software and software as a service. Eur. J. Oper. Res. **196**(2), 661–671 (2009)
P. Ghasemi, H.A. Khalili, A.P. Chobar, S. Safavi, F.M. Hejri, A new multiechelon mathematical modeling for pre-and postdisaster blood supply chain: robust optimization approach. Discret. Dyn. Nat. Soc. **2022**(1), 2976929 (2022)
P. Ghasemi, F. Goodarzian, V. Simic, E.B. Tirkolaee, A DEA-based simulation-optimisation approach to design a resilience plasma supply chain network: a case study of the COVID-19 outbreak. Int. J. Syst. Sci.: Operat. Logist. **10**(1), 2224105 (2023)
M. Giannakis, T. Papadopoulos, Supply chain sustainability: a risk management approach. Int. J. Prod. Econ. **171**, 455–470 (2016)
F. Goodarzian, V. Kumar, P. Ghasemi, A set of efficient heuristics and meta-heuristics to solve a multi-objective pharmaceutical supply chain network. Comput. Ind. Eng. **158**, 107389 (2021)
F. Goodarzian, P. Ghasemi, A. Appolloni, I. Ali, L.E. Cárdenas-Barrón, Supply chain network design based on big data analytics: Heuristic-simulation method in a pharmaceutical case study. Prod. Plann. Control 1–21 (2024)

F. Goodarzian, A. Abraham, J. Muñuzuri, A. Taghipour, P. Ghasemi, A benders decomposition approach for a new sustainable pharmaceutical supply chain network: a case study in France. Annals of Operat. Res. 1–39 (2024)

R. Goss, C. Maxim, D. Adhikari, J. Rothe, Leveraging new SEMI standard to reduce waste and improve flow for semiconductor manufacturing. Robot. Comput.-Integrat. Manufact. **26**(6), 658–664 (2010)

IBE Electronics, Top 10 global wafer foundries in 2023 (2024). https://www.pcbaaa.com/top-10-global-wafer-foundries/

D. Ivanov, *Introduction to Supply Chain Analytics With Examples in AnyLogic and anyLogistix Software* (Springer, 2024)

P.-X. Jiang, *Localization Analysis of Semiconductor Supply Chain* (2024). Master Degree Dissertation, National Yang Ming Chiao Tung University.

K.G. Kempf, Control-oriented approaches to supply chain management in semiconductor manufacturing. in *Proceedings of the 2004 American Control Conference* 5, (2004), pp. 4563–4576

V. Kumar, S. Srinivasan, A review of supply chain management using multi-agent system. Int. J. Comput. Sci. Issues **7**(5), 198 (2010)

I.W.G. Kwon, S.H. Kim, D.G. Martin, Healthcare supply chain management; strategic areas for quality and financial improvement. Technol. Forecast. Soc. Chang. **113**, 422–428 (2016)

H.K. Lam, Doing good across organizational boundaries: sustainable supply chain practices and firms' financial risk. Int. J. Oper. Prod. Manag. **38**(12), 2389–2412 (2018)

D.M. Lambert, *Supply Chain Management: Processes, Partnerships, Performance* (Supply Chain Management Inst., 2008)

Y.H. Lee, S. Chung, B. Lee, K.H. Kang, Supply chain model for the semiconductor industry in consideration of manufacturing characteristics. Prod. Plann. Control **17**(5), 518–533 (2006)

Y.-C. Lin, T. Chen, An advanced fuzzy collaborative intelligence approach for estimating the unit cost by fitting the underlying learning process. Complex and Intell. Syst. **5**, 303–313 (2019a)

C.-W. Lin, T. Chen, 3D printing technologies for enhancing the sustainability of an aircraft manufacturing or MRO company—A multi-expert partial-consensus FAHP analysis. Int. J. Adv. Manuf. Technol. **105**, 4171–4180 (2019b)

C.L. Martins, M.V. Pato, Supply chain sustainability: a tertiary literature review. J. Clean. Prod. **225**, 995–1016 (2019)

N. Matinrad, E. Roghanian, Z. Rasi, Supply chain network optimization: a review of classification, models, solution techniques and future research. Uncertain Supply Chain Managem. **1**(1), 1–24 (2013)

L. Mönch, R. Uzsoy, J.W. Fowler, A survey of semiconductor supply chain models part I: semiconductor supply chains, strategic network design, and supply chain simulation. Int. J. Prod. Res. **56**(13), 4524–4545 (2018a)

L. Mönch, C.F. Chien, S. Dauzère-Pérès, H. Ehm, J.W. Fowler, Modelling and analysis of semiconductor supply chains. Int. J. Prod. Res. **56**(13), 4521–4523 (2018b)

L. Peters, Reducing risk in the semiconductor supply chain (2024). https://semiengineering.com/reducing-risk-in-the-semiconductor-supply-chain/

A. Polater, O. Demirdogen, An investigation of healthcare supply chain management and patient responsiveness: an application on public hospitals. Int. J. Pharmaceut. Healthcare Market. **12**(3), 325–347 (2018)

A. Ross, N. Chen, E.Z. Hang, E.L. Glassman, F. Doshi-Velez, Evaluating the interpretability of generative models by interactive reconstruction. in *Proceedings of the 2021 CHI Conference on Human Factors in Computing Systems* (2021), pp. 1–15

A. Shilov, TSMC completes construction of 5nm fab 21 in Arizona (2022). https://www.tomshardware.com/news/tsmc-fab-21-arizona

A. Shilov, Samsung foundry: New $17 billion fab in the USA by late 2023 (2021). https://www.anandtech.com/show/16483/samsung-in-the-usa-a-17-billion-usd-fab-by-late-2023

Smart Electronics Industry Project Promotion Office, 2017 worldwide IC packaging and testing leaders (2018). https://www.sipo.org.tw/en/industry-overview/industry-ranking/ic-packaging-and-testing-industry-ranking.html

S. Somapa, M. Cools, W. Dullaert, Characterizing supply chain visibility–a literature review. The Int. J. Logist. Managem. **29**(1), 308–339 (2018)

Taipei Exchange, Introduction to the semiconductor industry chain (2024). https://ic.tpex.org.tw/introduce.php?ic=D000&stk_code=1410

TechNews, Semiconductor leaders cannot invest in the United States without the chemical industry Taiwan team (2020). https://technews.tw/2020/06/29/tsmc-big-business-opportunities-for-semiconductor-materials-in-taiwan/

The Directorate General of Budget, Accounting and Statistics, The total export value of integrated circuits from January to November 2011 was US$169.2 billion (2022). https://ws.dgbas.gov.tw/Download.ashx?u=LzAwMS9VcGxvYWQvNDYzL3JlbGZpbGUvMTEwMjAvMjMwNTgwLzgzY2NhMGZkLTM0MjUtNGM0NC1hOTM5LTA5YzYwM2M4OWM2OS5wZGY%3D&n=bjExMjAxMDQucGRm&icon=.pdf

The Economist, Taiwan's dominance of the chip industry makes it more important (2023). https://www.economist.com/special-report/2023/03/06/taiwans-dominance-of-the-chip-industry-makes-it-more-important

J. Toikka, The difference between tier 1, 2 and 3 suppliers (2023). https://sievo.com/blog/supplier-tiers

TrendForce, [News] China pushes local semiconductor supply chain, taiwanese companies reportedly see limited impact (2024). https://www.trendforce.com/news/2024/05/23/news-china-pushes-local-semiconductor-supply-chain-taiwanese-companies-reportedly-see-limited-impact/

M. Treppner, H. Binder, M. Hess, Interpretable generative deep learning: an illustration with single cell gene expression data. Hum. Genet. **141**(9), 1481–1498 (2022)

TSMC, A look at semiconductor supply chains (2024). https://www.tsmc.com/english/aboutTSMC/dc_infographics_supplychain

University Wafer, What are the wafer manufacturing process? (2024). https://www.universitywafer.com/wafer-manufacturing.html

Y.C. Wang, T. Chen, New XAI tools for selecting suitable 3D printing facilities in ubiquitous manufacturing. Complex and Intell. Syst. **9**(6), 6813–6829 (2023)

W. Wang, D.E. Rivera, Model predictive control for tactical decision-making in semiconductor manufacturing supply chain management. IEEE Trans. Control Syst. Technol. **16**(5), 841–855 (2008)

W. Wang, D.E. Rivera, K.G. Kempf, Model predictive control strategies for supply chain management in semiconductor manufacturing. Int. J. Prod. Econ. **107**(1), 56–77 (2007)

Y.C. Wang, T.C.T. Chen, M.C. Chiu, An improved explainable artificial intelligence tool in healthcare for hospital recommendation. Healthcare Anal. **3**, 100147 (2023a)

Y.C. Wang, T.C.T. Chen, Y.C. Lin, 3D printer selection for aircraft component manufacturing using a nonlinear FGM and dependency-considered fuzzy VIKOR approach. Aerospace **10**(7), 591 (2023b)

Y.C. Wang, T. Chen, M.C. Chiu, An explainable deep-learning approach for job cycle time prediction. Decision Anal. J. **6**, 100153 (2023c)

Y.C. Wang, T. Chen, Adapted techniques of explainable artificial intelligence for explaining genetic algorithms on the example of job scheduling. Expert Syst. Appl. **237**(A), 121369 (2024)

WikiChip, Chip-on-wafer-on-substrate (CoWoS)—TSMC (2024). https://en.wikichip.org/wiki/tsmc/cowos

C.H. Wu, OEM product design in a price competition with remanufactured product. Omega **41**(2), 287–298 (2013)

Z. Wu, F. Jia, Toward a theory of supply chain fields–understanding the institutional process of supply chain localization. J. Oper. Manag. **58**, 27–41 (2018)

H.-C. Wu, T. Chen, C.-H. Huang, A piecewise linear FGM approach for efficient and accurate FAHP analysis: smart backpack design as an example. Mathematics **8**, 1319 (2020a)

H.-C. Wu, T.-C.T. Chen, C.-H. Huang, Y.-C. Shi, Comparing built-in power banks for a smart backpack design using an auto-weighting fuzzy-weighted-intersection FAHP approach. Mathematics **8**(10), 1759 (2020b)

H.-C. Wu, T. Chen, M.-C. Chiu, Assessing the sustainability of smart healthcare applications using a multi-perspective fuzzy comprehensive evaluation approach. Digital Health **9**, 2055207623120390 (2023)

H.C. Wu, T.C.T. Chen, S.Y. Wang, Impact of the COVID-19 pandemic on the revenue of the catering industry: Taiwan as an example. SAGE Open **14**(2), 21582440241241410 (2024)

K.J. Wu, *The Origin of Specialization in Wafer Manufacturing in the Semiconductor Industry* (2014). Master Degree Dissertation, National Central University.

D. Zhang, Z. Sheng, J. Du, S. Jin, A study of emergency management of supply chain under supply disruption. Neural Comput. Appl. **24**, 13–20 (2014)

Chapter 2
Localization of Semiconductor Supply Chains—Driving Forces and Challenges

Abstract This chapter first examines the stability of a semiconductor supply chain based on two theories: the triple bottom line and the multi-generational philosophy. The sustainability of a semiconductor supply chain is clearly integral to its stability. To pursue the sustainability of a semiconductor supply chain, localization is considered a tactical strategy for foundries. The current localization trend is then compared to the traditional one to show its uniqueness. Subsequently, various forces driving the localization of semiconductor supply chains were identified as the U.S.-China trade war, geopolitics, the coronavirus disease 2019 (COVID-19) pandemic, the Ukraine-Russia war, environmental awareness, and the need for semiconductor supply chain transparency. The impact of each driving force is then described individually. In addition, several challenges that need to be overcome in the localization of semiconductor supply chains are also listed, including different laws, cultural differences, insufficient local human resources, uncertain local government subsidies, competition from local competitors, and higher wage levels. The procedure for semiconductor supply chain localization is subsequently established as an iterative process consisting of two parallel parts. Finally, possible future developments in semiconductor supply chain localization are discussed, with the impact of generative artificial intelligence (GenAI) and 3D packaging (e.g., chip-on-wafer-on-substrate, CoWoS) being also mentioned.

Keywords Localization · Semiconductor supply chain · Driving force · Generative AI · Geopolitics

2.1 Supply Chain Stability and Supply Chain Localization

Semiconductor supply chains use supply chain localization to pursue sustainability (Shilov 2023), which is closely related to the stability of supply chains (Wu et al. 2023). For this reason, supply chain stability is first defined.

Supply chain stability can be examined from two theories, namely Edgington's "**Triple Bottom Line**" (Elkington 2018) and the **multi-generational philosophy**

T. T. Chen, *Supply Chain Localization in the Semiconductor Industry*,
https://doi.org/10.1007/978-3-031-81280-4_2

(from the "Brundtland Report") (Burns and Witoszek 2012). Figure 2.1 illustrates the triple bottom line (Jiang 2024), which describes the phenomenon that companies aimed at maximizing **profits** in the past, but now they need to consider the **sustainability** of operations. Companies must take on three responsibilities: environmental responsibility, social responsibility, and economic responsibility. The multigenerational philosophy asserts that companies no longer prioritize profits, but must be environmentally friendly, meaning that they should meet current needs without compromising their ability to meet needs in the future. This is another definition of **sustainability**.

In order to maintain the stability of a supply chain, the supply chain must remain flexible, gain community trust, and adopt a localized procurement strategy. This is closely related to the development of **supply chain localization**, which can be integrated with Industry 4.0 (including cyber-physical systems, Internet of things (IoT), cloud computing and cognitive computing) to make information transparent (Chen 2024, 2023a; Wang and Chen 2023; Chen and Wang 2019), share information more accurately, and reduce unnecessary costs in the overall supply chain (see Fig. 2.2).

Fig. 2.1 Triple bottom line

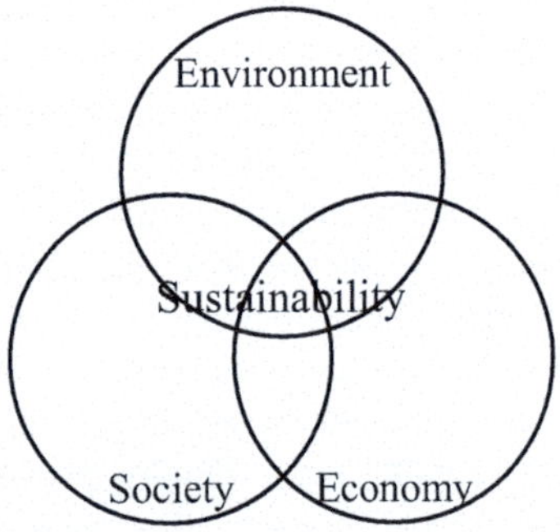

Fig. 2.2 Relationship among supply chain localization, stability and sustainability

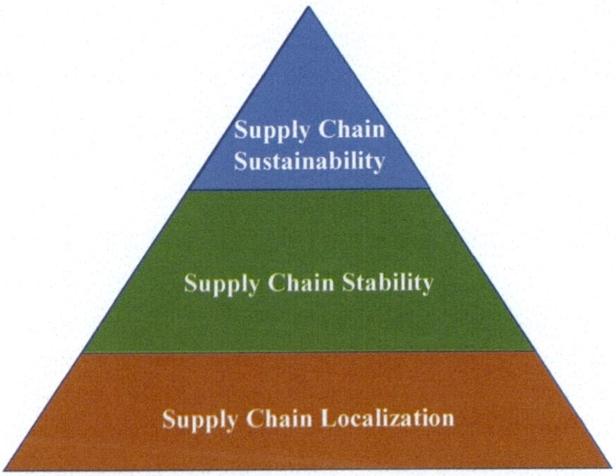

Supply chains in many industries have a trend of localized production. The conventional definition of supply chain localization is that **multinational enterprises produce and distribute products or provide services in the country (or region) where the customer is located** (that is, locally), rather than in the home country (i.e., domestically) (Wu and Jia 2018). In this way both factories and customers are located at the farthest end of the supply chain. This way is related to the circular (or sharing) economy, in which sharing circular resources in a region is believed to help maintain the stability of local supply chains (Sarkis 2020; Chen and Lin 2023; Chiu and Chen 2022).

The global demand for semiconductors is increasing day by day. The stability and convenience of semiconductor supply chains have become important issues. The localization of a semiconductor supply chain is closely related to its convenience and stability. Qualified raw materials and equipment to support semiconductor manufacturing come from different countries. In the past, semiconductor supply chain localization meant that wafer foundries attempted to rely on domestic suppliers for raw materials and/or equipment. Take Taiwan Semiconductor Manufacturing Co. (hereinafter referred to as TSMC), the chip foundry with the largest market share in the world, as an example. TSMC plays an important role in stabilizing semiconductor supply chains. To this end, TSMC is committed to diversifying risks to ensure the stability of chip supply (Strong 2023). To avoid the interruption of upstream raw materials and equipment, TSMC has been committed to local procurement in recent years by relying on domestic suppliers of special gas such as siliconethane (TechNews 2020).

However, there are other ways to implement supply chain localization. For example, recently supply chain localization has emerged in the semiconductor industry, in which **foundries (midstream of the supply chain) set up factories in the countries (or regions) where chip designers (upstream) are located** (Chen et al. 2023). Although chip designers are not the end consumers of chips, they are essentially one of foundries' buyers. For semiconductor supply chain localization, the top foundries, such as TSMC and Samsung, have gone to the United States with upstream manufacturers, such as suppliers of process chemicals, special gases, etc. TSMC has also negotiated conditions for these upstream suppliers, allowing them to enjoy the same land, subsidies and other preferential treatments as TSMC (TechNews 2020).

2.2 Localization of Semiconductor Supply Chains—Driving Forces

Global demand for semiconductors is surging day by day. As a result, the development of semiconductor supply chains has gradually become the focus of global attention. Semiconductor suppliers are spread across many countries. In recent years, the global industrial environment has changed dramatically, which has affected the stability and flexibility of semiconductor supply chains. As a result, semiconductor supply chains have shown a trend of localized development (Chen et al. 2023). In response to this trend, countries around the world have proposed corresponding measures for the semiconductor industry, such as enacting relevant laws to encourage the establishment of wafer fabs, or striving to develop local semiconductor supply chains to prevent over-reliance on foreign chip suppliers. In order to promote local semiconductor supply chains,

- The United States has enacted the **Chip Act** to encourage the investment in localized wafer fabs by foreign semiconductor companies (Kelly 2022).
- Japan and many European countries have also adopted national policies to attract investment in fabs to promote the localization of semiconductor supply chains (Chiang 2024; Blanchard and Escritt 2023).
- The EU, seeking to increase its strategic autonomy, has set out plans to double its market share of world semiconductor manufacturing from 10% in 2020 to 20% in 2030 (Rogers 2021).
- As part of a broader "Make in India" initiative, the Indian government has allocated $1 billion to subsidize companies setting up semiconductor manufacturing facilities in the country (Rogers 2021).

The stability of supply chains is highly valued by the semiconductor industry. The vertical integration of the upstream and downstream of semiconductor supply chains requires the cooperation between multiple companies, since the process from raw materials to chips must go through them. In the past, most of the upstream and downstream manufacturers in semiconductor supply chains were located around the world, forming a "**long chain**." Through the age of globalization, semiconductor supply chains distinguish the advantages of different regions in carrying out R&D, reducing production costs, and improving production efficiency (Chen et al. 2024a) when sitting facilities. The manufacturing of semiconductors usually requires long-distance transportation and the cooperation among manufacturers around the world. However, in recent years, supply chains in various industries no longer focus on long chains only, but have adopted a strategy of **long and short chains in parallel** (Trade Magazine 2021), and the same is true for semiconductor supply chains. Semiconductor supply chains no longer carry out global distribution only, but also emphasize localized production and move the industry back to their home countries. This phenomenon can prevent high risks caused by exports and avoid chain break crises. **Driving forces** for the localization of semiconductor supply chains in recent years are as follows (Fig. 2.3):

- **U.S.-China trade war;**
- **Geopolitics;**
- **The coronavirus disease 2019 (COVID-19) pandemic;**
- **Ukrainian-Russian war;**
- **Environmental awareness;**
- **Need for transparency in semiconductor supply chains** (Rogers 2021; Sodhi and Tang 2019).

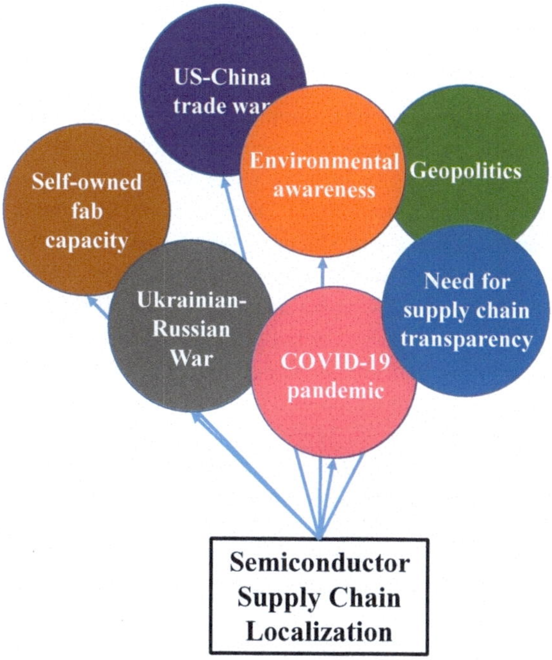

Fig. 2.3 Driving forces for semiconductor supply chain localization

as detailed in the following sections. In addition, some chip design companies are trying to own their own fabs to ensure production capacity and increase profits.

2.2.1 U.S.-China Trade War

The first driving force for the localization of semiconductor supply chains is **the U.S.-China trade war**, which also impacted the semiconductor manufacturing industry. For example, the United States seeks to limit or delay China's ability to acquire and produce advanced semiconductor manufacturing technologies (Shivakumar et al. 2024). In response to this, Nvidia plans to begin mass production of downgraded artificial intelligence (AI) chips it designs for China in the second quarter of 2024 to comply with U.S. export rules (Mo and Potkin 2024). The Dutch government has reached an agreement with the U.S. government to introduce a licensing system. If Advanced Semiconductor Materials Lithography (ASML) wants to export the most advanced lithography equipment to China, it must apply in advance to limit China's ability to make its own chips. As a consequence, ASML can continue to ship deep ultraviolet (DUV) lithography equipment-related products to China. However, DUV equipment is ASML's second-line product, and the company's most advanced extreme ultraviolet (EUV) lithography equipment has never been sold to China customers (Koc et al. 2024). In addition, if companies receive subsidies from the

Chip Act, they will be prohibited from establishing new fabs in China and Russia with wafer fabrication processes more advanced than 28 nm within ten years (Strong 2023).

In response to the economic protection of U.S. and China, various countries have formulated corresponding economic policies. Tariff barriers in economic policies, as well as the rise in labor costs in China, have led to the relocation of various industries (Maihold 2022). Due to this, supply chains become less stable, which has also increased the production cost of semiconductors, further pushing companies to set up factories in other places, thus promoting the trend of localization.

Semiconductor manufacturers adopt different localization strategies. Unlike TSMC, which moved or built its wafer fabrication capacity outside of Taiwan due to geopolitics, Samsung increased the proportion of its wafer fabs in South Korea while reducing that in China due to the U.S.-China trade war (Chiang and Hsiao 2023). For the same reason, Samsung also started to build a new wafer fab in US. It is worth noting that while TSMC is building new localized wafer fabs in the United States and Japan, it also continues to build up more wafer fabrication capacity in Taiwan, since continuing with existing practices has proven to be the most efficient and economical, albeit with some risks. Short-term competitiveness is still more important (or urgent) than long-term sustainability (Chen et al. 2024b, 2024c, 2024d; Lin and Chen 2019; Chen 2023b).

2.2.2 Geopolitics

During the U.S.-China trade war, both geographical and political factors have greatly affected the economic development of several countries. **Geopolitics** becomes an important factor causing the localization of semiconductor supply chains. In this regard, various countries have proposed “friendly shore outsourcing” policies (Chiang and Hsiao 2023), in order to keep core technologies domestically and hand over the rest to countries with similar political stances. This trade model can prevent over-reliance on countries with different political stances and ensure the stability of semiconductor supply chains.

The U.S. government tracks the concentration of various goods to prevent companies from relying on specific geographic locations and thus exposing themselves to geographic risk (Rogers 2021). U.S. semiconductor manufacturers began to expand their production scales to eliminate such risks, which also led to the localization of semiconductor supply chains.

Taiwan is located among many countries and is adjacent to China. It has a good geographical location. However, under the influence of geopolitics, Taiwan’s semiconductor supply chains tend to choose more stable regions to set up factories to ensure competitive advantages and avoid political risks. For example, TSMC planned to build two and three new wafer fabs in the United States and Japan, respectively. In the era of globalization, TSMC usually chose to build wafer fabrication capacity in China because China has low human resource costs and high education levels. TSMC

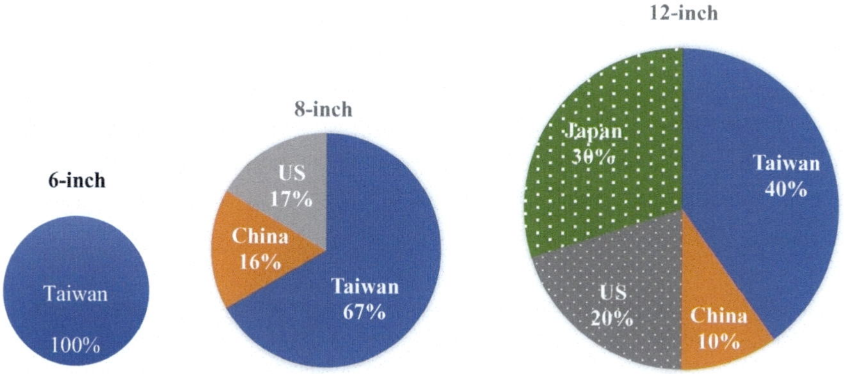

Fig. 2.4 TSMC's existing and planned wafer fab locations (*Data source* TSMC)

currently has two wafer fabs there (one 8-inch and one 12-inch) (see Fig. 2.4). The same advantages are no longer absolute in the United States and Japan. The main problem will be geopolitics. In Fig. 2.4, the dotted sectors represent the planned capacity. TSMC's localization plan is also presented in Table 2.1. A comparison of TSMC's fabs in Japan and the United States is shown in Fig. 2.5. A comparison of these two wafer fab localization plans is as follows:

- TSMC's wafer fab localization plan in Kumamoto was much smoother than that in Arizona. Subsidies from the Japanese government were earlier and more certain. The construction process was also smoother than that of the Arizona wafer fab.
- TSMC's first wafer fab in Kumamoto started mass production earlier than that in Arizona.
- TSMC has expressed its intention to build at least three fabs in these two countries.
- Because the needs of customers in the two countries are different, the wafer fabrication technologies used by the fabs are also different. The fabs in Arizona are more advanced than those in Kumamoto. However, this situation may change in the future.

Another country where TSMC is building localized wafer fabs is Germany (that is, the European Union). TSMC will invest 20 billion euros in Dresden wafer fab and expect to start construction in the second half of 2024, which is basically in line with the schedule announced by TSMC last year (Alkousaa and Mukherjee 2023). The joint venture partners of this wafer fab localization plan include German giants Bosch, Infineon and Dutch semiconductor company NXP. EU and German government subsidies account for about half of the investment in the localized wafer fab.

By comparison, Samsung has four production centers in South Korea, one in the United States, and three in China. Samsung has a much higher proportion of production centers in China than TSMC (see Fig. 2.6). Therefore, for Samsung, the impact of the U.S.-China trade war is greater and the motivation for semiconductor

Table 2.1 TSMC's localization plan (*Data source* TSMC)

Time	Event
April, 2022	Japan government provided $3 billion financial support
	TSMC began the construction of its first wafer fab in Kumamoto, Japan
May, 2020	TSMC announced the construction of the first wafer fab in Arizona, U.S
	TSMC planned to invest $12 billion
December, 2022	TSMC announced the construction of the second wafer fab in Arizona, U.S
	TSMC increased the total investment to $40 billion
Februry, 2024	TSMC announced the construction of the second wafer fab in Japan, also in Kumamoto
	Japan government increased the financial support to $10 billion
Februry 24, 2024	TSMC opened its first wafer fab in Japan
April, 2024	U.S. government decided to provide TSMC with $6.6 billion in direct funding
	TSMC announced the construction of the third wafer fab in Arizona, U.S
	TSMC increased the total investment to $65 billion
	TSMC started the trial production at the first wafer fab in U.S
December, 2024	TSMC plans to start the production at its first wafer fab in Japan
	TSMC plans to start the construction of the second wafer fab in Japan
First half, 2025	TSMC plans to start the mass production at its first wafer fab in U.S
December, 2027	TSMC plans to start the production at its second wafer fab in Japan
2028	TSMC plans to start the mass production at its second wafer fab in U.S

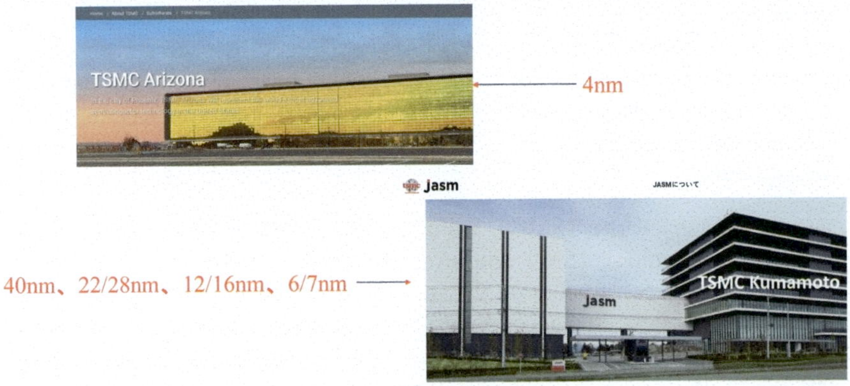

Fig. 2.5 Comparison of TSMC's fabs in Japan and the United States (*Data source* TSMC)

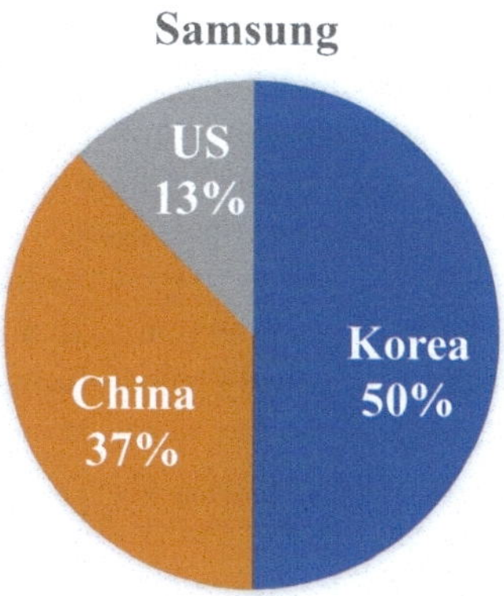

Fig. 2.6 Samsung's wafer fab locations (*Data source* Samsung)

supply chain localization is stronger. However, Samsung can easily avoid this impact by setting up new wafer fabs domestically, while TSMC is still subject to geopolitics.

As mentioned earlier, Samsung's localization strategy is to increase the capacity of wafer fabs in South Korea, which is motivated by the U.S.-China trade war. If geopolitics is taken into consideration instead, Samsung may build new fabs outside the Korean peninsula to avoid the risk of war.

2.2.3 *COVID-19 Pandemic*

The COVID-19 pandemic broke out globally at the beginning of 2020. Moosavi et al. (2022) pointed out that supply chains were interrupted due to the outbreak, which affected all industries around the world especially small and medium-sized enterprises. It has been particularly dramatic, with production halted as employees were unable to work and customer demand was reduced (Chen 2023c; Wu et al. 2024; Chen and Wu 2023; Lin and Chen 2022).

As the pandemic raged on, the outputs from the upstream of supply chains have decreased, resulting in a shortage of parts in the overall supply chain. For example, the main production site for disinfectant spray nozzles is in China, one of the first areas where the cases of the COVID-19 pandemic were detected (Kilpatrick 2020). As a result, the manufacturers of disinfectant sprayers cannot assemble and ship them in a short time. Automakers have also been forced to postpone vehicle deliveries as part suppliers temporarily closed to slow the spread of COVID-19 (Howard 2021).

Global supply chains were therefore facing a crisis of disconnection. The COVID-19 pandemic has forced some suppliers to close. As a result, the contracts with these suppliers were canceled, breached or ignored. Buyers have no choice but to look for alternative suppliers (Keegan 2020). Therefore, how to select suitable alternative suppliers for manufacturers has become a critical task during the COVID-19 pandemic (Supply and Demand Chain Executive 2020; Chen et al. 2021). Nevertheless, supplier relationships have encountered unprecedented challenges (Weissman

2020). Howorth (Weissman 2020) believed that supplier relationships will transform from short-term trade nature to long-term collaboration and strategic relationships.

In addition, the competition for the limited capacity of alternative suppliers will intensify in the upstream or midstream segments of supply chains (Howorth 2020). To explore this topic, Ivanov (2020) conducted a simulation study to assess and predict the impact of the COVID-19 pandemic on global supply chains. Hoek (2020) analyzed seven companies that adopted **total cost of ownership**, **supplier segmentation**, and **supply chain change management theory** to mitigate the impact of the COVID-19 pandemic. Based on the experimental results, finding alternative suppliers became increasingly important to reduce supply chain risks. Sharma et al. (2022) developed a framework for enhancing supply chain survivability during and after the COVID-19 pandemic by optimizing six criteria: **performance under uncertainty**, **configuration**, **governance works**, **viability**, **collaboration**, and **digital data-driven**. Majumdar et al. (2020) used a case to highlight the vulnerability of apparel supply chains in South Asian countries during the COVID-19 pandemic. They then recommended that disruption risk-sharing agreements be included in procurement contracts and that suppliers be prohibited from subcontracting without authorization.

User motivation and acceptance of many products varied during the COVID-19 pandemic. As a result, some supply chains can be sustain while others cannot. In addition, global events during the COVID-19 pandemic, including the U.S.-China trade war, severe delays and congestion at ports and terminals due to the COVID-19 pandemic, global semiconductor chip shortages, inflation, and the Ukraine-Russia war, were also having an impact on supply chains sustainability. To address this issue, the impact of these global events on supply chain resilience needs to be discussed. To this end, Chen (2023d) re-evaluated the sustainability of smart device supply chains from objective and subjective perspectives based on the evidence collected during the COVID-19 pandemic.

Ivanov and Dolgui (2020) discussed the completeness and feasibility of intertwined supply networks composed of companies playing different roles (buyers or suppliers) in different supply chains. It was believed that **intertwined supply networks** can serve society more sustainably than other types of supply chains during the COVID-19 pandemic. Wang and Chen (2021) asserted that buyers may hold multiple viewpoints simultaneously regarding the relative priorities of criteria for selecting suitable alternative suppliers during the COVID-19 pandemic. To address this issue, they proposed a bi-objective analytic hierarchy process (AHP)–mixed integer-nonlinear programming (MINLP)–genetic algorithm (GA) approach. First, arithmetic mean was applied to decompose the buyer's judgment matrix into several sub-judgment matrices. Each sub-judgment matrix represented a single viewpoint and generated a priority set using AHP. To generate diversified priority sets, a bi-objective MINLP problem was solved using a GA, and multiple alternative suppliers could be selected based on these priority sets. For the same purpose. Chen (2021) proposed a diversified AHP-tree approach instead, in which sub-judgment matrices were diversified by maximizing their distances from each other.

However, the impact of a pandemic on supply chains was not limited to the short term. There will also be long-term impacts on enterprises. Response measures taken by companies also need to be prioritized (Magableh and Mistarihi 2022).

2.2.4 *Ukrainian-Russian War*

New forms of supply chains will focus on the localization of upstream and downstream enterprises. For example, the U.S. semiconductor industry hopes to localize wafer fabrication to establish U.S. semiconductor supply chains. When Ukraine-Russia war broke out in 2022, in addition to overcoming the impact of the COVID-19 pandemic, various industries also had to withstand the negative impact of the war. Russia and Ukraine produce a number of raw materials that are needed by countries around the world. The raging artillery fire has significantly reduced the supply of many key raw materials. For semiconductor supply chains, the raw material inert gases (such as neon and argon) required for the production of chips mostly come from Ukraine and Russia. According to the statistics in Chiao (2022),

- **Neon gas** produced by Russia and Ukraine accounts for 70% of the global market. Neon is present in the air at only 0.00182% and is one of the raw materials for excimer laser gas used in semiconductor exposure processes.
- 90% of **inert gases** required by the U.S. semiconductor industry come from Ukraine.

After Russia's annexation of Crimea in 2014, the global neon price increased by approximately 600%, prompting some chip manufacturers to begin switching from Russian and Ukrainian suppliers to Chinese suppliers. The Ukraine-Russian war in 2022 also forced two Ukrainian companies to shut down, which was expected to exacerbate the chip shortage caused by the COVID-19 pandemic (DeCarlo et al. 2022; Chen et al. 2024e; Wang et al. 2023a; Chen 2023e). Semiconductor supply chains will take a hit as Ukraine-Russia war continues.

As of 2022, Ukraine remains the primary source of semiconductor-grade neon imported into the United States. The price of neon has tripled as China closed steel plants and reduced demand for Australian iron ore to reduce pollution. The return to normal supply of neon depends on when the Ukraine-Russia war ends, how quickly Ukrainian production can restart, and when Ukrainian neon can re-enter the global market (Aju News 2022).

In 2022, a Japanese company successfully localized the production equipment and technology for inert gases "neon" used in semiconductors (Wu et al. 2022). In addition, at least one U.S. company expanded domestic neon production (Aju News 2022). These actions made foreign foundries (such a TSMC and Samsung) establish localized wafer fabs in these countries a niche.

2.2.5 Environmental Awareness

In addition to political and pandemic factors, environmental issues have always been taken seriously by the semiconductor industry. In the past, based on comparative interests, upstream and downstream manufacturers in supply chains were mostly distributed in countries around the world, and the delivery of products relied on air and sea freight. However, in order to reduce **industrial pollution** and **carbon emissions**, countries around the world are promoting zero carbon emissions, and the localized production of supply chains was emphasized in the study of Wu et al. (Frost et al. 2023) as being helpful in reducing carbon emissions.

Take TSMC, the chip foundry with the largest market share in the world, as an example. The company is committed to upgrading its local supply chain and reducing its carbon footprint through the local procurement of indirect raw materials. In order to localize suppliers in line with its **low-carbon policy**, the company manages and actually visits its suppliers. According to the company's supply chain localization management policy, it conducts data management for specific suppliers to facilitate long-term tracking. Suppliers use the Supply Online 360 system to manage data and report to TSMC the progress of achieving goals, allowing TSMC to know the status of each supplier. At the same time, it also provides relevant education and training courses to educate suppliers on the concept of **carbon reduction**. The Supply Online 360 system has been used since 2020. According to its official website, TSMC applies this supply chain management platform to sustainable operations. Suppliers can also use this platform to **self-rate** themselves and **learn online** courses to maintain the sustainability of supply chains digitally and make supply chains operate more efficiently. Figure 2.7 shows the four major policies of TSMC's low-carbon supply chain management. However, such a system seems to do little to increase the two-way transparency in the supply chain, as it only provides/manage suppliers' details to TSMC in one direction.

2.2.6 Need for Semiconductor Supply Chain Transparency

Corporate transparency means that a company discloses information to consumers, investors and other stakeholders about its supply chain operations and products' compliance with consumer expectations (Sodhi and Tang 2019).

The COVID-19 pandemic has reduced air and sea freight capacity as workers become infected, leaving many industries, including semiconductor manufacturing, short of raw materials. Supply chain disruptions during this period have severely impacted the automotive industry due to a shortage of automotive chips. They were forced to cut production and reduce vehicle supply (Shivakumar et al. 2024). Semiconductor supply constraints also impacted automakers' ability to produce enough aftermarket repair parts to keep existing vehicles on the road. Therefore, semiconductor supply chains, especially their transparency, must be subject to a rigorous

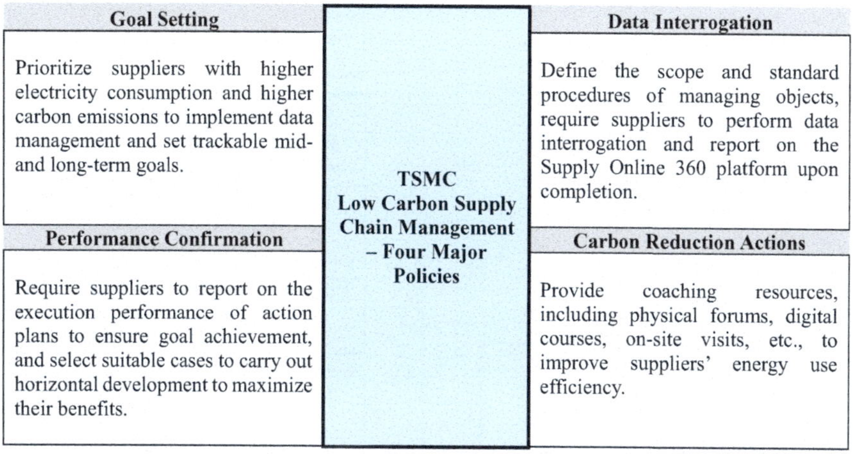

Fig. 2.7 Four major policies of TSMC's low-carbon supply chain management (*Data source* TSMC)

review. Localizing the semiconductor supply chain by building new wafer fabs locally will apparently enable chip designers to effectively oversee foundries and take timely measures when things go wrong, thereby increasing supply chain transparency.

2.3 Challenges to Overcome in Semiconductor Supply Chain Localization

According to the experience of some foundries, the key tasks of semiconductor supply chain localization are as follows (see Fig. 2.8):

- Select the most suitable city (location) to build the localized wafer fabrication capacity;
- Strive for financial subsidies from the central government;
- Negotiate tax incentives and exemptions with the local government;
- Ensure the timely availability of resources required for the construction of the new wafer fab;
- Ensure worker safety during the construction process of the new wafer fab;
- Migrate domestic workers and in-service equipment to the new wafer fab;
- Go to local colleges and universities to advertise the foundry and the new wafer fab;
- Recruit and train local workers.

However, semiconductor companies around the world provide different services and products to the market, and their localization strategies are also different. For example, in order to expand its scale and pursue sustainable development, TSMC

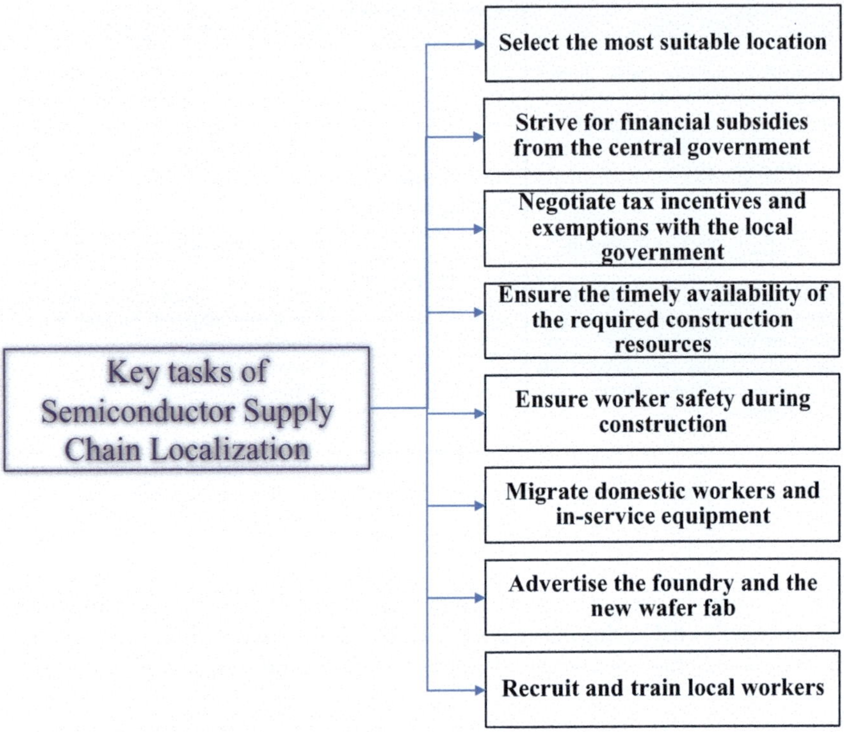

Fig. 2.8 Key tasks of semiconductor supply chain localization

has actively set up wafer fabs overseas. However, the **costs and benefits** of setting up wafer fabs in a different country are difficult to evaluate. There are also other challenges to overcome, such as **different laws**, **cultural differences**, **insufficient local human resources**, **uncertain local government subsidies**, **competition among local competitors**, **higher wage levels**, etc. (see Fig. 2.9). Nevertheless, building new wafer fabrication capacity in locations not previously considered has become an important issue. Morris Chang, the founder of TSMC, the world's largest wafer foundry manufacturer, even said that "In the chip sector, globalization is dead." (Frost et al. 2023).

Intel announced two years ago that it would spend 30 billion euros to build a new localized wafer fab in Magdeburg, eastern Germany, but it has not yet determined a date for the start of construction (Alkousaa and Mukherjee 2023). Intel's slow construction of new localized wafer fabs in Germany may be related to the company's poor financial performance. In the face of financial difficulties, Intel's overseas investment plans tend to be conservative. For similar reasons, American semiconductor manufacturer Wolfspeed's plan to build a localized wafer fab in the Saarland in southwestern Germany has also been significantly delayed. The reason was that the company hoped to obtain more subsidies from the German government.

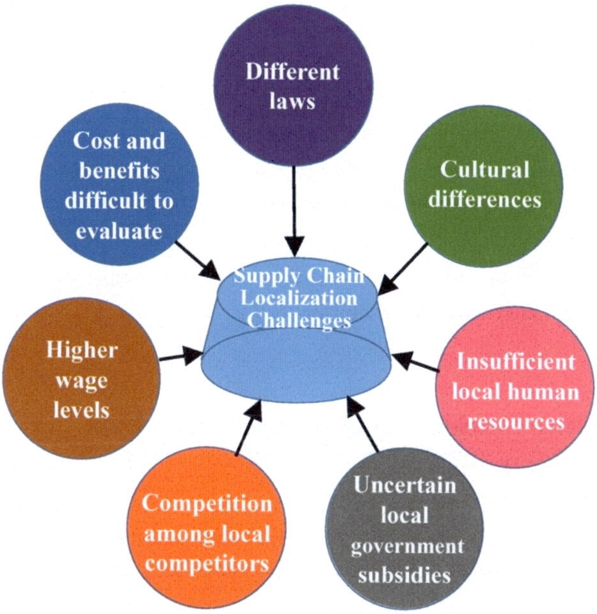

Fig. 2.9 Challenges to overcome in semiconductor supply chain localization

2.4 Prioritizing Possible Measures to Overcome the Challenges

Possible measures that can be taken by these semiconductor manufacturers in overcoming supply chain localization challenges include:

- Benefiting local residents or leveraging local communities to alleviate cultural differences;
- Designing staggered shift systems;
- Cross-shift collaboration in problem solving;
- Collaboration between wafer fabs in different time zones;
- Increasing (or relocating) the proportion of domestic workers;
- Further automation;
- Accelerating yield learning through early mass production;
- Etc.

These possible measures were to be prioritized. In other words, the purpose of conducting this experiment was not to select the best possible measure. Foundries are certainly and willing to take multiple measures simultaneously. Chen et al. (2023) considered conformity to local needs, local government support and subsidies, total investment required, closeness to third parties, and low political, war, and other risks in choosing the location of a new wafer fab for semiconductor supply chain localization. However, in this case, the wafer fab location has been determined so these factors no longer needed to be determined. Instead, the following factors were considered:

- Implementation costs;
- Legal compliance;
- Easiness to implement;
- Need for cross-organizational collaboration;
- Quick and visible results.

Implementation costs and the need for cross-organizational collaboration are STB criteria, while the others are LTB criteria.

2.5 Procedure for Semiconductor Supply Chain Localization

The procedure for semiconductor supply chain localization is illustrated in Fig. 2.10, which is an iterative process consisting of two parallel parts (Chen and Lin 2023; Chen 2023f; Wang et al. 2023b):

(Part I)

Step I-1. Deciding on the country (or region) to set up a new localized wafer fab.
Step I-2. Negotiate with the government of the country (or region) where the new localized wafer fab is located to seek more policy support and financial subsidies.
Step I-3. Select the city (or location) to build the new localized wafer fab.
Step I-4. Control the building process of the new localized wafer fab.
Step I-5. Conduct trial production at the new localized wafer fab.
Step I-6. Conduct mass production at the new localized wafer fab.

(Part II)

Step II-1. Continuously monitoring competitors' localization actions.
Step II-2. If a competitor decides to build a new localized wafer fab at the same location, initiate another wafer fab localization plan; otherwise, return to Step II-1.

2.6 Future Development of Semiconductor Supply Chain Localization

Some future developments in semiconductor supply chain localization are predicted in this section. First, some of the forces driving the localization of semiconductor supply chains are disappearing, such as **the COVID-19 pandemic**. The resumption of free cross-border transportation reduces the risk of semiconductor supply chain disruptions and alleviates the urgency of supply chain localization. What may change

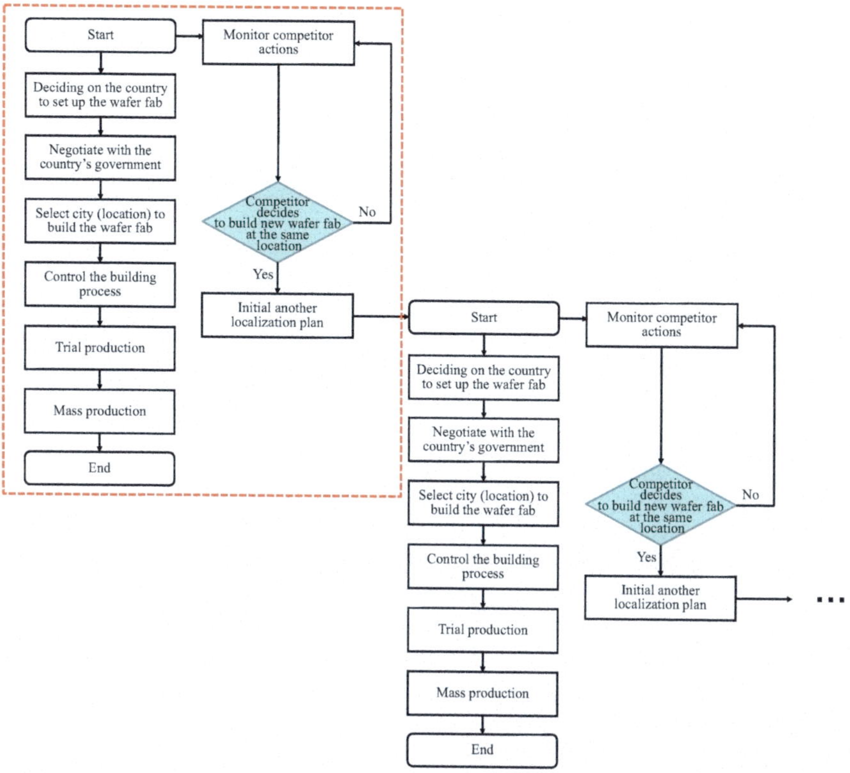

Fig. 2.10 Procedure for semiconductor supply chain localization

next is the Ukrainian-Russian War and the U.S.-China trade war. Geopolitics and other driving forces are not easy to disappear in the short term.

Secondly, with the popularity of **generative artificial intelligence (GenAI)** represented by ChatGPT, there is an increasing demand for chips, services, workstations and even personal computers (PCs) that support such applications. Most of the AI chips used for high-speed computing are manufactured by TSMC. In addition, the U.S.-China trade war and geopolitics have made it less likely that other hardware will be produced in other places such as China. Therefore, in the AI wave, not only wafer foundries, but Taiwan's PC and server supply chains are also relied on, which weakens TSMC's urgency to build new wafer fabs abroad.

Chip-on-wafer-on-substrate (CoWoS) is TSMC's two-point five-dimensional integrated circuit (2.5D IC) packaging technology based on a through-silicon via (TSV) interposer designed for high-performance applications (WikiChip and (CoWoS)—TSMC 2024; Chen 2023g). Since advanced packaging has reached below 7 nm, the R&D speed of traditional packaging facilities can no longer keep up with the pace of the wafer fabrication process. The CoW part of CoWoS is too precise and can only be manufactured by TSMC, which has caused this situation. At the same

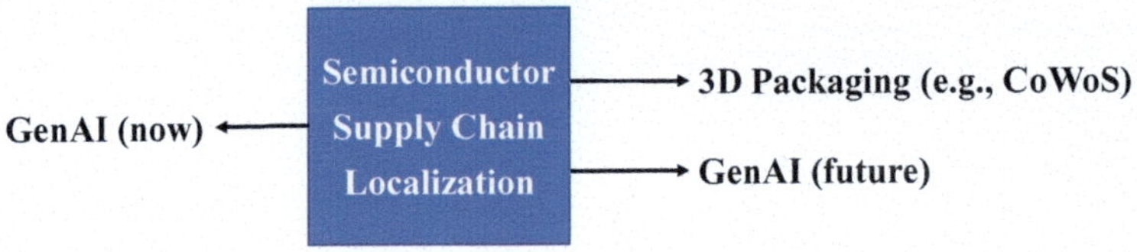

Fig. 2.11 Impact of GenAI and 3D packaging (e.g., CoWoS) on semiconductor supply chain localization

time, TSMC has many customers around the world, especially in the United States, who need such high-end products. Therefore, its "one-stop" service can maintain the yields of wafer fabrication and packaging at the same time (Wang and Chen 2024). The emergence of CoWoS has enhanced TSMC's independence from packaging facilities to a certain extent and is conducive to TSMC's attempts to localize its semiconductor supply chain (see Fig. 2.11).

However, the further development of localization of the semiconductor supply chain obviously depends on whether these foundries have good and successful experiences in building localized wafer fabs, and how these localized wafer fabs perform in the future, which will take time to evaluate. For example, relying on the successful experience of building the first localized wafer fab in Kumamoto and the timely and generous support of the Japanese government, TSMC announced the construction of the second and even the third localized wafer fab near Kumamoto (Chen and Wang 2024; TSMC 2024).

In addition, several major chipmakers, including ASML, Intel and its suppliers, Applied Materials and Nova, already have fabs in Arizona. It is intuitive for wafer foundries to follow the experiences of others in making such decisions (Chen 2023h; Nenni 2024). In short, successful attempts will further promote the development of localization of semiconductor supply chains. From another perspective, foundries that have suffered short-term failures in localizing fabrication capabilities may poach employees from successful rivals by offering higher wages if they are willing to wait.

2.6.1 IDM 2.0 vs. Foundry 2.0

As one of the world's largest integrated device manufacturers (IDMs), Intel is trying to expand its foundry business and "all-in" all available resources to find trustworthy alternatives to TSMC for foundry services, so it proposed IDM 2.0 concept (Chen 2023i). In contrast, TSMC, one of the world's largest foundry manufacturers, is gradually expanding the functional items included in its foundry through its own downstream capabilities in semiconductor manufacturing such as 3D packaging, becoming the so-called foundry 2.0 (Scharf et al. 2021; Chen and Wang 2022; Wang et al. 2023c, 2023d; Chen 2023j; Chen et al. 2020).

In the future, more downstream and even upstream functional items may be included in foundry 2.0. For TSMC, it will be able to gradually get rid of its dependence on third parties. However, TSMC will also become a larger monopoly (monster) in the semiconductor industry and may be subject to more pressure from all parties, posing challenges to its sustainability.

References

Aju News, POSCO succeeds in domestically producing equipment and technology for the production of rare gas "neon" for semiconductors. (2022). https://japan.ajunews.com/view/20220113115013262

R. Alkousaa, S. Mukherjee, Germany earmarks 20 billion euros in subsidies for chip industry (2023). https://www.reuters.com/technology/germany-earmarks-20-bln-eur-chip-industry-coming-years-2023-07-25/

B. Blanchard, T. Escritt, Germany spends big to win $11 billion TSMC chip plant (2023). https://www.reuters.com/technology/taiwan-chipmaker-tsmc-approves-38-bln-germany-factory-plan-2023-08-08/

T.R. Burns, N. Witoszek, Brundtland report revisited: toward a new humanist agenda. J. Hum. Ecol. **39**(2), 155–170 (2012)

T. Chen, A diversified AHP-tree approach for multiple-criteria supplier selection. CMS **18**(4), 431–453 (2021)

T. Chen, C.-W. Lin, Optimizing 3D printing facility selection for ubiquitous manufacturing using an evolving fuzzy big data analytics approach. Int. J. Adv. Manuf. Technol. **127**, 4111–4121 (2023)

T. Chen, Y.-C. Wang, An advanced IoT system for assisting ubiquitous manufacturing with 3D printing. Int. J. Adv. Manuf. Technol. **103**, 1721–1733 (2019)

T.C.T. Chen, Y.C. Wang, Evaluating innovative future robotic applications in manufacturing using a fuzzy collaborative intelligence approach. The Int. J. Adv. Manuf. Technol. **130**(11), 6027–6041 (2024)

T. Chen, H.-C. Wu, A partial-consensus and unequal-authority fuzzy collaborative intelligence approach for assessing robotic applications amid the COVID-19 pandemic. Soft. Comput. **27**, 16493–16509 (2023)

T. Chen, Y.C. Wang, H.C. Wu, Analyzing the impact of vaccine availability on alternative supplier selection amid the COVID-19 pandemic: a cFGM-FTOPSIS-FWI approach. Healthcare **9**(1), 71 (2021)

T. Chen, Y.C. Wang, P.H. Jiang, A selectively calibrated derivation technique and generalized fuzzy TOPSIS for semiconductor supply chain localization assessment. Decision Anal. J. **8**, 100275 (2023)

T. Chen, C.-W. Lin, Y.-C. Lin, A fuzzy collaborative forecasting approach based on XAI applications for cycle time range estimation. Appl. Soft Comput. **151**, 111122 (2024a)

T.C.T. Chen, H.C. Wu, K.W. Hsu, Recommending suitable hotels to travelers in the post-COVID-19 pandemic using a novel FAHP-fuzzy TOPSIS approach. Complex and Intell. Syst. **10**, 6901–6915 (2024b)

T.C.T. Chen, H.C. Wu, M.C. Chiu, A deep neural network with modified random forest incremental interpretation approach for diagnosing diabetes in smart healthcare. Appl. Soft Comput. **152**, 111183 (2024e)

T.C.T. Chen, Industry 4.0 for semiconductor manufacturing. in *Production Planning and Control in Semiconductor Manufacturing: Big Data Analytics and Industry 4.0 Applications* (2023), pp. 21–40

T.C.T. Chen, Enhancing the sustainability of smart healthcare applications with XAI. in *Sustainable Smart Healthcare: Lessons Learned from the COVID-19 Pandemic* (2023), pp. 93–110

T.C.T. Chen, Smart technology applications in healthcare before, during, and after the COVID-19 pandemic. in *Sustainable Smart Healthcare: Lessons Learned from the COVID-19 Pandemic* (2023), pp. 19–37

T.C.T. Chen, Sustainable smart healthcare applications: lessons learned from the COVID-19 pandemic. in *Sustainable Smart Healthcare: Lessons Learned from the COVID-19 Pandemic* (2023), pp. 65–92

T.C.T. Chen, Smart healthcare. in *Sustainable Smart Healthcare: Lessons Learned from the COVID-19 Pandemic* (2023), pp. 1–18

T.C.T. Chen, Applications of XAI for decision making in the manufacturing domain. in *Explainable Artificial Intelligence (XAI) in Manufacturing: Methodology, Tools, and Applications* (2023), pp. 51–81

T.C.T. Chen, Big data analytics for semiconductor manufacturing. in *Production Planning and Control in Semiconductor Manufacturing: Big Data Analytics and Industry 4.0 Applications* (2023), pp. 1–19

T.C.T. Chen, Applications of XAI for forecasting in the manufacturing domain. in *Explainable Artificial Intelligence (XAI) in Manufacturing: Methodology, Tools, and Applications* (2023), pp. 13–50

T.C.T. Chen, Applications of XAI for forecasting in the manufacturing domain. in *Explainable Artificial Intelligence (XAI) in Manufacturing: Methodology, Tools, and Applications* (2023), pp. 1–11

T.C.T. Chen, Applications of XAI to job sequencing and scheduling in manufacturing. in *Explainable Artificial Intelligence (XAI) in Manufacturing: Methodology, Tools, and Applications* (2023), pp. 83–105

T.C.T. Chen, XAmI applications to telemedicine and telecare. in *Explainable Ambient Intelligence (XAmI) Explainable Artificial Intelligence Applications in Smart Life* (2024), pp. 85–90

T. Chen, Y.C. Wang, Hybrid big data analytics and industry 4.0 approach to projecting cycle time ranges. The Int. J. Adv. Manuf. Technol. **120**(1), 279–295 (2022)

T.C.T. Chen, W.A. Chaovalitwongse, M.J. O'grady, K. Honda, Editorial: smart technologies for improving the quality of mobile health care. Health Care Managem. Sci. **23**, 171–172 (2020)

T.C.T. Chen, Y.C. Wang, C. W. Lin, Enhancing the effectiveness of output projection in wafer fabrication using an industry 4.0 and XAI approach. The Int. J. Adv. Manuf. Technol. **134**(1), 113–125 (2024)

T. Chen, Y.-C. Wang, Y.-C. Lin, Production planning and control of a pharmaceutical supply chain using evolutionary fuzzy approach. J. Eng. Res. 1–10 (2024)

S. Chiang, Chip giant TSMC opens first Japan factory as it diversifies away from Taiwan amid U.S.-China tensions (2024). https://www.cnbc.com/2024/02/26/tsmc-opens-new-plant-in-japan-as-it-diversifies-away-from-taiwan.html

D. Chiang, J. Hsiao, Samsung localizes supplier base amid rising geopolitical risks (2023). https://www.digitimes.com/news/a20230823PD200/samsung-supplier-south-korea-ict-manufacturing.html

J. Chiao, Ukrainian-Russian conflict affects semiconductor gas supply and may cause rise in chip production costs, says TrendForce. (2022). https://www.trendforce.com/presscenter/news/20220215-11119.html

M.-C. Chiu, T. Chen, A ubiquitous healthcare system of 3D printing facilities for making dentures: application of type-II fuzzy logic. Digital Health **8**, 20552076221092540 (2022)

S. DeCarlo, S. Goodman, O Ukraine, neon, and semiconductors. (2022). https://www.usitc.gov/publications/332/executive_briefings/ebot_decarlo_goodman_ukraine_neon_and_semiconductors.pdf

J. Elkington, 25 years ago I coined the phrase "Triple Bottom Line." Here's why it's time to rethink it (2018). https://hbr.org/2018/06/25-years-ago-i-coined-the-phrase-triple-bottom-line-heres-why-im-giving-up-on-it#:~:text=About%2025%20years%20ago%2C%20John%20Elkington%20coined%20the,lives%20people%20and%20the%20health%20of%20the%20planet

E. Frost, K. Liu, TSMC founder: in the chip sector, globalization is dead. (2023). https://english.cw.com.tw/article/article.action?id=3396

R.V. Hoek, Responding to COVID-19 supply chain risks—Insights from supply chain change management, total cost of ownership and supplier segmentation theory. Logistics **4**, 23 (2020)

P.W. Howard, Ford Bronco deliveries delayed until summer because of COVID-19 supply chain disruptions (2021). https://www.usatoday.com/story/money/cars/2020/12/05/2021-ford-bronco-coronavirus-delays-summer/3840894001/

D. Howorth, Will COVID-19 spark a new approach to retailer/supplier relationships? (2020). https://www.foodmanufacture.co.uk/Article/2020/11/13/Will-COVID-19-spark-a-new-approach-to-retailer-supplier-relationships

D. Ivanov, Predicting the impacts of epidemic outbreaks on global supply chains: a simulation-based analysis on the coronavirus outbreak (COVID-19/SARS-CoV-2) case. Transp. Res. E Logist. Transp. Rev. **136**, 101922 (2020)

D. Ivanov, A. Dolgui, Viability of intertwined supply networks: extending the supply chain resilience angles towards survivability. A position paper motivated by COVID-19 outbreak. Int. J. Prod. Res. **58**, 2904–2915 (2020)

P.-X. Jiang, *Localization Analysis of Semiconductor Supply Chain* (2024). Master Degree Dissertation, National Yang Ming Chiao Tung University.

K. Keegan, COVID-19: operations and supply chain disruption (2020). https://www.pwc.com/us/en/library/covid-19/supply-chain.html

M. Kelly, CHIPS Act clears Congress, ensuring $52 billion boost to US foundries (2022). https://www.theverge.com/2022/7/28/23282494/semiconductors-chips-and-science-act-joe-biden-congress-nvidia

J. Kilpatrick, COVID-19: Managing supply chain risk and disruption (2020). https://www2.deloitte.com/global/en/pages/risk/articles/covid-19-managing-supply-chain-risk-and-disruption.html

C. Koc, I. King, D. Baazil, ASML's China sales surged despite secret Dutch deal with US (2024). https://www.bloomberg.com/news/articles/2024-01-25/asml-s-china-sales-surged-despite-secret-dutch-deal-with-us

C.-W. Lin, T. Chen, 3D printing technologies for enhancing the sustainability of an aircraft manufacturing or MRO company—A multi-expert partial-consensus FAHP analysis. Int. J. Adv. Manuf. Technol. **105**, 4171–4180 (2019)

Y.C. Lin, T.C.T. Chen, Type-II fuzzy approach with explainable artificial intelligence for nature-based leisure travel destination selection amid the COVID-19 pandemic. Digital Health **8**, 20552076221106320 (2022)

G.M. Magableh, M.Z. Mistarihi, Applications of MCDM approach (ANP-TOPSIS) to evaluate supply chain solutions in the context of COVID-19. Heliyon **8**(3), e09062 (2022)

G. Maihold, A new geopolitics of supply chains: The rise of friend-shoring (2022). https://www.ssoar.info/ssoar/handle/document/81700

A. Majumdar, M. Shaw, S.K. Sinha, COVID-19 debunks the myth of socially sustainable supply chain: a case of the clothing industry in South Asian countries. Sustain. Prod. Consum. **24**, 150–155 (2020)

Y. Mo, F. Potkin, Nvidia to launch China-focused AI chip in Q2 2024—sources (2024). https://www.reuters.com/technology/nvidia-launch-china-focused-ai-chip-q2-2024-sources-2024-01-08/

J. Moosavi, A.M. Fathollahi-Fard, M.A. Dulebenets, Supply chain disruption during the COVID-19 pandemic: recognizing potential disruption management strategies. Int. J. Disaster Risk Reduct. **75**(1), 102983 (2022)

D. Nenni, TSMC foundry 2.0 and Intel IDM 2.0 (2024). https://semiwiki.com/semiconductor-manufacturers/tsmc/347646-tsmc-foundry-2-0-and-intel-idm-2-0/

C. Rogers, Everyone loves localization of semiconductors; Applied Materials, Intel act (2021). https://www.spglobal.com/marketintelligence/en/news-insights/latest-news-headlines/everyone-loves-localization-of-semiconductors-applied-materials-intel-act-63661421

J. Sarkis, Supply chain sustainability: learning from the COVID-19 pandemic. Int. J. Oper. Prod. Manag. **41**(1), 63–73 (2020)

S. Scharf, B. Sander, M. Kujath, H. Richter, E. Riedel, H. Stein, J.T. Felde, Foundry 4.0: an innovative technology for sustainable and flexible process design in foundries. Proc. CIRP **98**, 73–78 (2021)

M. Sharma, S. Luthra, S. Joshi, A. Kumar, Developing a framework for enhancing survivability of sustainable supply chains during and post-COVID-19 pandemic. Int J Log Res Appl **25**(4–5), 433–453 (2022)

A. Shilov, Samsung foundry: New $17 billion fab in the USA by late 2023 (2021). https://www.anandtech.com/show/16483/samsung-in-the-usa-a-17-billion-usd-fab-by-

S. Shivakumar, C. Wessner, T. Howell, Balancing the ledger: Export controls on U.S. chip technology to China (2024). https://www.csis.org/analysis/balancing-ledger-export-controls-us-chip-technology-china

M.S. Sodhi, C.S. Tang, Research opportunities in supply chain transparency. Prod. Oper. Manag. **28**(12), 2946–2959 (2019)

M. Strong, TSMC sees Taiwan semiconductor business as element of regional stability (2023). https://www.taiwannews.com.tw/news/4911416

Supply and Demand Chain Executive, Companies pursue alternative suppliers to spread supply chain risks and build resilience to mitigate COVID-19 impact (2020). https://www.sdcexec.com/risk-compliance/press-release/21203708/companies-pursue-alternative-suppliers-to-spread-supply-chain-risks-and-build-resilience-to-mitigate-covid19-impact

TechNews, Semiconductor leaders cannot invest in the United States without the chemical industry Taiwan team (2020). https://technews.tw/2020/06/29/tsmc-big-business-opportunities-for-semiconductor-materials-in-taiwan/

Trade Magazine, In the reorganization of the global industrial supply chain, the impact of the trade war and the pandemic has changed the "long chain" into a "short chain" (2021). https://www.ieatpe.org.tw/magazine/ebook356/storypage08.html

TSMC, JASM set to expand in Kumamoto Japan. (2024). https://pr.tsmc.com/english/news/3105

Y.C. Wang, T. Chen, A bi-objective AHP-MINLP-GA approach for flexible alternative supplier selection amid the COVID-19 pandemic. Soft Comput. Lett. **3**, 100016 (2021)

Y.C. Wang, T. Chen, New XAI tools for selecting suitable 3D printing facilities in ubiquitous manufacturing. Complex and Intell. Syst. **9**, 6813–6829 (2023)

Y.C. Wang, T.C.T. Chen, M.C. Chiu, An improved explainable artificial intelligence tool in healthcare for hospital recommendation. Healthcare Anal. **3**, 100147 (2023a)

Y.C. Wang, T. Chen, M.C. Chiu, An explainable deep-learning approach for job cycle time prediction. Decision Anal. J. **6**, 100153 (2023b)

Y.C. Wang, T.C.T. Chen, H.C. Wu, A novel auto-weighting deep-learning fuzzy collaborative intelligence approach. Decision Anal. J. **6**, 100186 (2023d)

Y.C. Wang, T. Chen, Adapted techniques of explainable artificial intelligence for explaining genetic algorithms on the example of job scheduling. Expert Syst. Appl. **237**(A), 121369 (2024)

Y.C. Wang, T.C.T. Chen, M.C. Chiu, A systematic approach to enhance the explainability of artificial intelligence in healthcare with application to diagnosis of diabetes. Healthcare Anal. **3**, 100183 (2023)

R. Weissman, Navigating supplier relationships in the COVID-19 era (2020). https://www.supplychaindive.com/news/coronavirus-supplier-contracts-relationships/576132/

WikiChip, Chip-on-wafer-on-substrate (CoWoS)—TSMC (2024). https://en.wikichip.org/wiki/tsmc/cowos

Z. Wu, F. Jia, Toward a theory of supply chain fields–understanding the institutional process of supply chain localization. J. Oper. Manag. **58**, 27–41 (2018)

H. Wu, H. Mehrabi, N. Naveed, P. Karagiannidis, Impact of strategic control and supply chain management on recycled plastic additive manufacturing. J. Clean. Prod. **364**, 132511 (2022)

H.-C. Wu, T. Chen, M.-C. Chiu, Assessing the sustainability of smart healthcare applications using a multi-perspective fuzzy comprehensive evaluation approach. Digital Health **9**, 2055207623120390 (2023)

H.-C. Wu, T. Chen, S.Y. Wang, Impact of the COVID-19 pandemic on the revenue of the catering industry: Taiwan as an example. SAGE Open **14**(2), 21582440241241410 (2024)

Chapter 3
Issues in Managing Localized Semiconductor Supply Chains

Abstract This chapter first emphasizes that localization is one of the important factors for the sustainable development of a supply chain. The advantages of localization mentioned in the supply chain management literature are then summarized. Several related issues in semiconductor supply chain management are also raised. The first is how to choose the most suitable factory location to localize a wafer fab. For comparison purposes, the literature on facility location selection in traditional supply chains is also reviewed. Several real cases are also analyzed. The analysis results show that the country (or region) chosen to build a localized wafer fab is actually irrelevant. Instead, the most suitable city or location in the intended country (or region) must be selected to build the localized wafer fab. The second issue is how to mitigate the impact of cultural differences. To this end, a number of measures have been proposed, including seeking the assistance from local communities, the cooperation between fabs in different time zones, the natural replacement of domestic workers, hiring more local employees, designing notification or reminder forms, and a staggered shift system. The next issue discusses the planning horizon of a semiconductor supply chain localization plan, for which the concept of flexible long-term localization is proposed. The final issue is about the wafer technology node that a localized wafer fab should adopt to meet the needs of local customers, foundries and local governments.

Keywords Semiconductor supply chain · Localization · Facility location selection · Cultural differences · Flexible long-term localization · Wafer technology node

3.1 Supply Chain Localization

Localization is one of the important factors in making supply chains sustainable. In response to the rise of environmental awareness, localization of supply chains can reduce carbon emissions. Wu et al. (2022) mentioned that localization can reduce carbon emissions and reduce the dependence on supply chains, thereby allowing

T. T. Chen, *Supply Chain Localization in the Semiconductor Industry*,
https://doi.org/10.1007/978-3-031-81280-4_3

the demand for labor and job opportunities to increase in the region. Koo and Kim (2022) explored the localization of the Korean electronics industry in Vietnam due to lower labor costs. This move made the scale and investment amount of the Korean electronics industry in Vietnam increase day by day, thereby procurement could be carried out locally. However, their study also pointed out that although Vietnam has sufficient labor force, there are still some gaps in the required technologies (Chen 2023a). A similar situation has occurred in the United States. Their semiconductor industry still lacks the technologies or experiences to fabricate wafers economically and with high quality, and therefore requires the building of localized wafer fabs (Texas Economic Development Corporation 2021).

Affected by the COVID-19 pandemic, global supply chain management has different development strategies. Due to the stagnation of suppliers' operations caused by the pandemic, the overall supply chain is in a disconnection crisis. Nandi et al. (2021) pointed out that the crisis caused by the COVID-19 pandemic has caused supply chains to adopt a localized strategy to reduce risks arising from distance. Localization can reduce tariffs, transportation and labor costs, and can also increase the agility of supply chains. Cherrafi et al. (2022) mentioned that localization of supply chains can help improve the flexibility of supply chains, and resources can be effectively used through the circular economy to reduce risks caused by the COVID-19 pandemic. Choi et al. (2021) pointed out that due to environmental influences, globalized supply chains will tend to use localized supplies. Localized supply chains are less complex and can use local resources flexibly. Frieske and Stieler (2022) pointed out that during the COVID-19 pandemic, semiconductor supply chain disruption crises also affected the development of other industries. Therefore, it was extremely important to maintain the stability of semiconductor supply chains. Local production and finding alternative suppliers are key to maintaining the stability of supply chains. Table 3.1 summarizes the advantages of localization mentioned in the supply chain management literature. Other advantages include being able to compete fairly with local competitors and bid for local government and military contracts (James 2021).

Table 3.1 Advantages of supply chain localization

Reference	Advantages of supply chain localization
Cherrafi et al. (2022)	Improve supply chain flexibility
Choi et al. (2021)	Be uncomplicated and able to use local resources
Frieske and Stieler (2022)	Prevent semiconductor supply chain disruptions and shutdowns
Koo and Kim (2022)	Procure locally to reduce costs
Nandi et al. (2021)	Reduce costs and improve supply chain agility
Wu et al. (2022)	Reduce carbon emissions and help protect the environment

3.2 Facility Location Selection

The first issue in managing a localized semiconductor supply chain is choosing a location to build a localized wafer fab, which is closely related to supplier selection. Therefore, before exploring this issue, the supplier selection problem and facility location selection problem in a traditional supply chain are first reviewed.

3.2.1 Supplier and Facility Location Selection for Traditional Supply Chains

Supply chain management requires making many choices, including the selection of upstream and downstream factory sites and suppliers. There are many factors to consider when selecting a supplier, and these factors may not fall into one category. For example, Kilincci et al. (2011) considered fourteen criteria for judging supplier suitability, divided into three categories: supplier performance, product performance, and service performance, which can be taken into account in choosing suitable suppliers.

In addition to supplier selection, facility location selection (or facility sitting) is also important to business growth. Kahraman et al. (2003) studied how automobile companies choose the best location to build factories. The criteria they considered were environmental regulations, community locations, competitive advantage and political risk. Many references consider multiple levels of criteria; each level contains several sub-criteria representing factory-set details. Appropriate facility siting can help stabilize a supply chain and have a long-term impact on the development of related companies and the efficiency of the entire supply chain.

The principles of sustainable business operations must consider social, environmental and economic dimensions. Therefore, various factors must be considered when selecting a facility location. When relocating a factory, planners not only target economic benefits, but also consider social and environmental responsibilities to maximize the overall welfare.

Therefore, in supply chain management, social, environmental and economic factors are taken into consideration when developing factory site evaluation criteria. For food industry supply chains, Mohseni et al. (2022) considered some criteria from three aspects (social, environmental and economic) and various sub-criteria under these criteria to assist in plant site selection. Kheybari et al. (2019) studied facility siting for renewable energy production facilities. They considered the setting of a factory site from three aspects: social, environmental and economic aspects, and established decision-making criteria for the three aspects. Table 3.2 shows a comparison of supply chain facility location considerations in the literature.

Table 3.2 Comparison of supply chain facility location considerations in the literature

Reference	Goal	Considered criteria
Biswas and Pamucar (2020)	Location selection of business school facilities	• Location awareness • Convenience of traveling • Driving time • Connectivity to markets • Connectivity to employers and industrial areas • Availability of rides • Availability of medical facilities • Internet accessibility • Environmentally friendliness • Security • Transportation costs • Familiarity with local language
Dogan (2012)	Global facility location selection for manufacturing companies	• Labor skills • Labor enthusiasm • Labor availability • Supplier capability • Number of alternative suppliers • Supplier reliability and influence • Population • Language proficiency • Unemployment rate • Land availability and price • Raw material quality and availability • Climate • Standard of living • Education system • Health system • Financial services • Credit reliability • Financial risk • Air and water transport quality • Land transportation quality • Infrastructure quality • Political stability • Bureaucratic culture • Government business legislation • Number of technical facilities • Number of scientific facilities • Amount of infrastructure • Environmental regulations • Tax structure • Tariffs • Cultural differences • Community attitude • Degree of unionization • Currency stability • Economic growth rate • Degree of inflation

(continued)

Table 3.2 (continued)

Reference	Goal	Considered criteria
Kahraman et al. (2003)	Factory location assessment for car companies	• Environmental regulations • Community • Competitive advantage • Political risk
Kheybari et al. (2019)	Renewable energy facility location selection	• Policy and legal support • Labor force • Acceptance • Impact on society • Ash management • Energy storage • Impact on resources and nature reserves • Greenhouse gas emissions • Energy saving • Distance from historical tourist areas • Greenhouse gas emissions • Investment cost • Transportation costs • Production and operation costs
Torkayesh and Simic (2022)	Medical plastic waste recycling facility site setting	• Flexibility • Land requirement • After-sales service • Average transportation cost • Average transportation distance • Traffic congestion level • Investment • Operation and maintenance costs • Financial support • Land degradation • Carbon footprint • Noise pollution • Degree of policy coordination • Social acceptance • Potential employment opportunity • Proportion of affected population

3.2.2 *Facility Location Selection for Localized Supply Chains*

Due to customer demand, wafer foundries are actively building localized wafer fabs in the countries where customers are located. For example, TSMC, the chip foundry with the largest market share in the world, is actively building new wafer fabs in the United States and Japan (James 2021; The Economist and is having more luck building in Japan than in America 2024), even though production costs in both countries are higher than that in Taiwan. In particular, the U.S. government and customers have exerted tremendous pressure to avoid geopolitics risks. In addition, the orders from U.S. customers currently account for TSMC's largest revenue. Therefore, TSMC

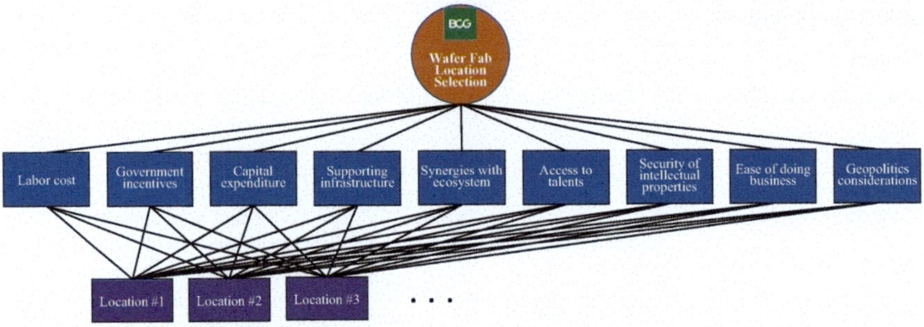

Fig. 3.1 Boston Consulting Group's factors for localized wafer fab location selection

does not need to consider which country or region to set up new factories. It started planning to set up a factory in the United States very early, although subsequent disputes and difficulties continued.

According to Boston Consulting Group (Varas et al. 2020), factors critical for selecting wafer fab locations include (see Fig. 3.1).

- Labor cost;
- Government incentives;
- Capital expenditure;
- Supporting infrastructure;
- Synergies with ecosystem/footprints;
- Access to talents;
- Security of intellectual properties (IP);
- Ease of doing business;
- Geopolitics considerations.

For localizing semiconductor supply chains,

- The choice of country (or region) to build a new fab becomes irrelevant. According to the estimations by the Boston Consulting Group, if the U.S. government provides $20 billion in subsidies, the United States can attract five more new wafer fabs. If subsidies were doubled, this number would increase to 10 (Varas et al. 2020). Such high subsidies and political pressure make other countries less comparable.
- Instead, the most suitable city or location in a predetermined country (or region) must be selected to build a new wafer fab. In this case, TSMC has chosen Arizona, where it planned to build a $3.5 billion wafer fab that the company said will create 1,600 new jobs and cost a total of $12 billion to build between 2021 and 2029. It is intuitive to follow the experiences of others in making such a decision, as several major chipmakers, including ASML, Intel and its suppliers, Applied Materials and Nova, already have fabs in Arizona (The Economist and is having more luck building in Japan than in America 2024). So Samsung, TSMC and NXP chose Arizona, the so-called silicon desert, over Austin (James 2021).

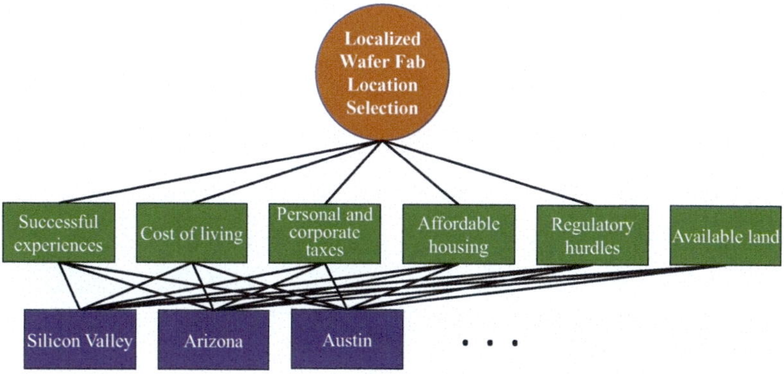

Fig. 3.2 Considerations in localized wafer fab location selection after the country was determined

The reason is simple—increasing expenses for Silicon Valley businesses and workers are driving people to move to more economical tech hubs, which is, obviously, Arizona, known for its lower cost of living, lower personal and corporate taxes, more affordable housing, available land, and fewer regulatory hurdles (see Fig. 3.2).

In comparison, the plan to build new factories in Japan is extremely smooth (The Economist and is having more luck building in Japan than in America 2024), and TSMC even plans to build three new wafer fabs there.

In addition, according to the documents submitted by Samsung, the world's largest semiconductor memory manufacturer, to Texas authorities, another way to select a localized wafer fab location is to consider four criteria that the location must meet to contribute to Samsung's leadership (i.e., competitiveness): talent acquisition, existing semiconductor manufacturing ecosystem systems, speed to market, and strong public–private partnerships (i.e., incentives) (see Fig. 3.3) (James 2021).

3.3 Difficulties in Supply Chain Localization

According to the Economist (The Economist and is having more luck building in Japan than in America 2024), truculent workers and red tape have slowed TSMC's efforts in Arizona. **Cultural differences**, **differences in laws between countries**, **insufficient qualified human resources**, **low worker compliance**, **lack of resources**, and **competition from (domestic) rivals** are all reasons for its failure in the United States (Fig. 3.4):

- These difficulties differ from the challenges described in Fig. 2.9 in that country-related considerations no longer matter.

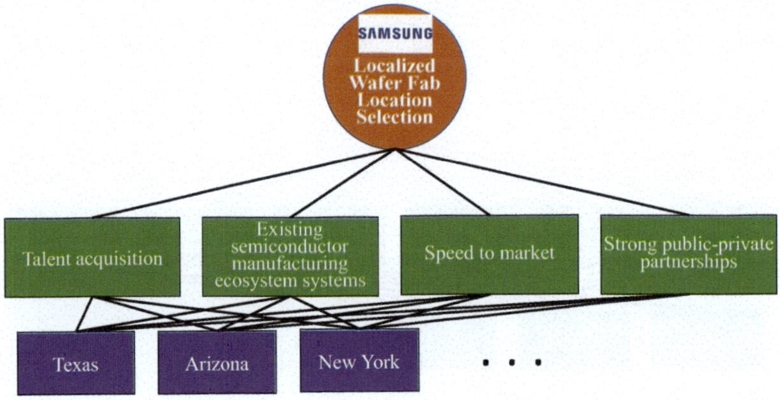

Fig. 3.3 Samsung's criteria for localized wafer fab location selection

Fig. 3.4 Difficulties faced by TSMC in supply chain localization

- In addition, most of these difficulties are unavoidable, but their impact can be mitigated to some extent with the help of local governments, communities, third parties, etc.

As a result, TSMC's new wafer fab in Arizona, USA, started construction one year earlier than that in Kumamoto, Japan, but it entered mass production one year later. Even if this had been more or less expected, TSMC still had to build another new fab in the United States.

3.3.1 Mitigating the Impact of Cultural Differences

There are two cultural differences in semiconductor supply chain localization. The first is owing to the transfer of domestic workers to the new localized wafer fab that leads to behavior differences between them and local workers. The second is due to modifying the existing practice of domestic wafer fabs to adapt to the new localized wafer fab (see Fig. 3.5).

For example, TSMC recently opened a new wafer fab in Kumamoto Prefecture, Japan. To facilitate the start-up of the new wafer fab, many existing wafer fabrication equipment used by wafer fabs in Taiwan were transferred to the new wafer fab. Some Taiwanese employees have also been transferred to work in the new wafer fab. Such a treatment is indeed necessary to maintain TSMC's short-term competitiveness (Varas et al. 2020; Chen 2023b, 2023c, 2023d; Yunogami 2006; Wang and Chen 2023; Solucis Santhapparaj et al. 2006; Chen et al. 2024a; Wang et al. 2023a; Wu et al. 2023). The new wafer fab will also employ more and more local Japanese to mitigate the impact of cultural differences (TSMC 2024). However, according to JobMenta (2023), there are still five cultural differences between the Japanese and Taiwanese workplaces:

- Vertical organization and culture of respecting seniors;
- Strictly adhering to the agenda during meetings, and focusing on reaching team consensus before making decisions;
- Emphasizing on teamwork and collaboration: Team success takes precedence over individual success;
- Being accustomed to using indirect expressions and non-verbal communication;
- Strictly requiring proper business etiquette and codes of conduct.

The impact of such cultural differences is to be mitigated with the help of local communities:

- **Diffusion effect**: Local organizations in Kumamoto Prefecture are actively recruiting non-semiconductor-related students from Taiwanese universities to work in Kumamoto. This is a diffusion effect and also helps to ease cultural differences (The Asahi Shimbun2023). In addition, TSMC's Kumamoto wafer

Fig. 3.5 Two cultural differences in semiconductor supply chain localization

fab projects were also expected to spur Taiwanese investment in Japan (Focus Taiwan 2023), being another diffusion effect.

- **Promoting cultural exchange**: Kumamoto tour packages were launched by local organizations to make employees from Taiwan and their families more familiar with Kumamoto's environment to reduce cultural shock (The Yomiuri Shimbun 2024).
- **Forming living circles** (The Asahi Shimbun2023).
- **Living facilities with domestic features**: Taiwan product counters were set up in Kumamoto department stores to provide services to Taiwanese employees, which also helps mitigate the impact of cultural differences (ETtoday 2024).

However, some cultural differences take time to ease. For example, when there is a problem in production, Taiwanese worker may choose to work overtime to solve the problem, while Japanese worker may decide to deal with it tomorrow (Business Weekly2024).

For a comparison, TSMC also established a new wafer fab in Arizona, USA. Some Taiwanese employees of the new wafer fab in Arizona, USA, complained that managers treat local employees better (Chen 2024). According to Chen 2023e, the cultural differences between the U.S. and Taiwanese workplaces include the following:

- More direct and informal communication;
- More informal culture, placing less emphasis on hierarchy and more on teamwork and collaboration;
- More casual and informal attitudes, not constrained by etiquette and manners;
- Focusing more on results than details;
- Tendency to prioritize current goals over future outcomes;
- Emphasizing individual achievement and competition rather than teamwork and harmony;
- Lower willingness to work overtime;
- Emphasizing work-life balance and personal time.

Compared with the Kumamoto case, the contribution of the local government or communities to mitigating cultural conflicts in the Arizona case is less obvious. As a result, due to the needs of domestic employees, a local living circle, dubbed "TSMC Village", has naturally formed near TSMC's Arizona wafer fab to reduce cultural shock (2022), with **living facilities** such as coffee shops and petrol stations nearby, as well as shopping centres, supermarkets, cinemas and restaurants. Some Taiwanese catering companies and Bank of Taiwan have also entered the city at the same time, further helping ease cultural conflicts. TSMC's Arizona wafer fab is expected to boost the local real estate market and bring a large number of employees and investors to Phoenix. These results **benefit local residents**, which may be most beneficial in alleviating cultural differences. A similar phenomenon can also be observed in TSMC's Kumamoto case. However, local residents in Kumamoto may not enjoy the increase in land and housing prices.

Table 3.3 Comparison of the actions taken by TSMC's two localized wafer fab plans in Kumamoto and Arizona to alleviate cultural differences

Plan	Diffusion effect	Promoting cultural exchange	Forming living circles	Living facilities with domestic features	Benefiting local residents,
TSMC's Kumamoto localized wafer fab	****	*****	*****	*****	***
TSMC's Arizona localized wafer fab	*****	*	*****	****	*****

A comparison of the actions taken by TSMC's two localized wafer fab plans in Kumamoto and Arizona to alleviate cultural differences is shown in Table 3.3. The number of asterisks is proportional to the effect.

3.3.2 Systemic Measures

If such cultural differences are not addressed, the performance of a localized wafer fab will inevitably be degraded. For example, when day shift workers discover a product quality (or machine reliability) problem, domestic workers tend to work overtime until the problem is fully or at least partially resolved, while local workers may leave the problem to be solved until the next day. This has nothing to do with the quality or diligence of workers in different regions, due to cultural or behavior differences.

One way to solve this problem is to design a staggered shift system, that is, add a shift in which workers work across day and swing shifts (half day shift and half swing shift) (i.e., the first new shift in Fig. 3.6). Such workers will be notified the product quality (or machine reliability) problem by day shift workers and inform swing shift workers to work together to solve the problem. If the product quality problem cannot be easily solved, a new shift (i.e., the second new shift in Fig. 3.6) can be added, spanning the swing and graveyard shifts, and continue to solve the product quality (or machine reliability) problem together with day shift workers. This staggered shift system was once common during the COVID-19 pandemic, as workers' schedules were constrained by many pandemic-related issues (Chen 2023f; Wu et al. 2024; Chen and Wu 2023), such as caring for infected family members or children at home because of school closures.

The second systematic measure to mitigate the impact of such problems is to design a notification or reminder form (see Fig. 3.7) that is given by day shift workers to night shift workers so that they can quickly understand the situation and try to resolve it. This handover form should include a description of the problem, when it

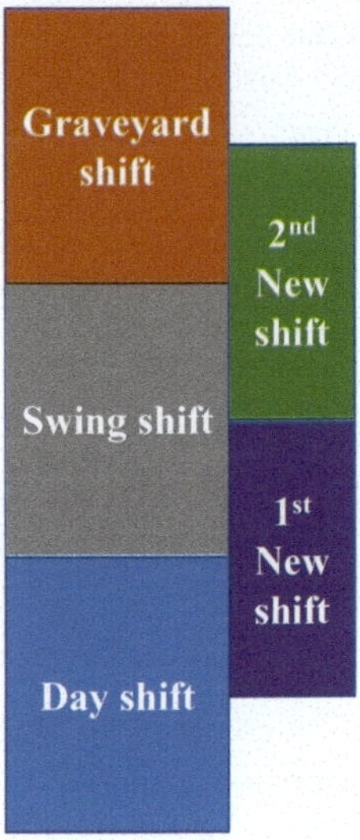

Fig. 3.6 Staggered shift system

occurred, possible causes, possible solutions, and at least actions that should be taken by night shift workers. Filling out and passing this handover form electronically is not really intended to resolve the problem, but rather to assist the night shift staff in quickly understanding the problem.

Another systemic approach to solving such problems is through the cooperation between fabs in different time zones (see Fig. 3.8). For example, when a product

Made by			
To			
Time of occurrence			
Problem Description		Picture	
Possible Causes			
Possible Solutions			
Actions to be Taken			

Fig. 3.7 Handover form

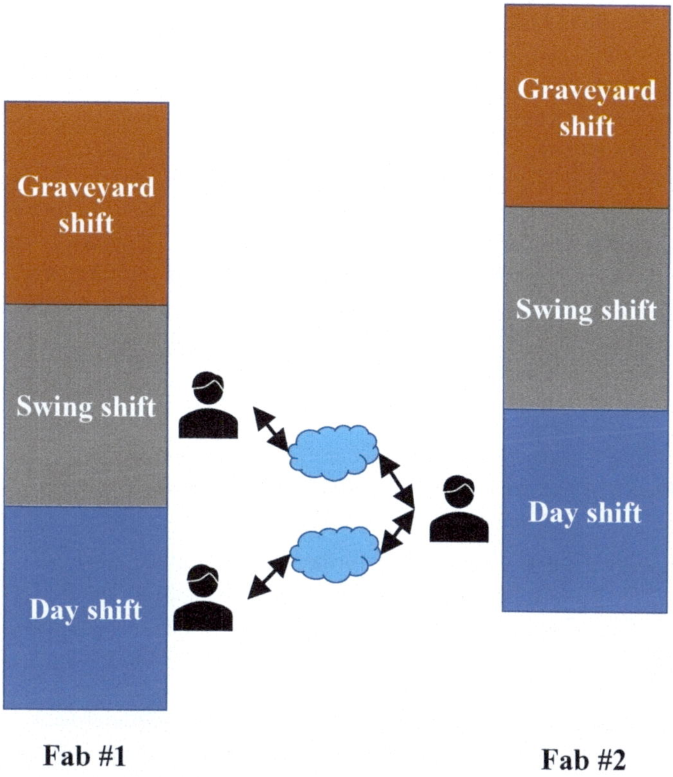

Fig. 3.8 Cooperation between fabs in different time zones

quality (or machine reliability) problem occur, engineers from fabs in different time zones are invited to join an online meeting to understand the problem and provide suggestions. Then, after a shift change, engineers on the night shift can get a quick update from engineers at the fab in a different time zone.

Wafer foundries transfer domestic workers to local wafer fabs to avoid slow production learning. As time goes by, more and more local workers will enter localized wafer fabs to replace these domestic workers, and cultural differences will naturally disappear. This is also a systematic measure.

3.3.3 *Implication for the Practice*

Such systematic measures to solving product quality (or machine reliability) problems is of particular practical interest, since human involvement is a prerequisite for the rapid resolution and future prevention of such problems. Take TSMC as an example (2024). In TSMC, machine learning (ML) is used to implement automatic defect classification (ADC) and maintain the accuracy of advanced defect classification (ADC) identification. The ML method used in TSMC is human–machine loop (HITL), which leverages human and machine intelligence to create ML models. HITL has been used in a variety of situations. HITL is related to modeling and simulation (M&S) in the immediate, virtual and constructive classifications.

With enough data and human tuning, the applied ML algorithms can quickly and extremely accurately identify and analyze images without the need for engineers to constantly teach machines what defects actually look like. Therefore, human intelligence will play an important role in ML, which means that the experience and algorithms provided are critical and unique to successful ML models. For these purposes, ensuring that those involved are aware of the situation is a must, and this is what the proposed measures aim for.

3.3.4 Making Good Use of Workers' Domain Knowledge

After mitigating cultural differences, a further step is to make good use of workers' domain knowledge to solve product quality, machine reliability or other production-related problems, which can also be seen in the literature. Since employees come from different backgrounds (i.e., a source of cultural differences), their domain knowledge is expected to complement each other.

Chen and Wang (2023) highlighted the problem that artificial neural networks (ANN) or deep neural networks (DNN) applications in manufacturing are usually difficult to understand and/or communicate. This problem hinders their acceptability (and practicality). Additionally, existing explainable artificial intelligence (XAI) techniques use simpler decision rules to approximate the prediction (or detection) process and/or result of an ANN (or DNN). However, these decision rules may not be consistent with domain knowledge, which limits their usefulness. They therefore proposed a modified random forest incremental interpretation (MRFII) method to solve this problem. MRFII includes the following steps:

Step 1. Apply the RF method to approximate the input-output relationship of the ANN (or DNN).
Step 2. Model domain knowledge (in terms of decision rules).
Step 3. For the target object (job, wafer, or machine), extract applicable decision rules from all decision trees.
Step 4. Evaluate the conformance of each decision rule to domain knowledge.
Step 5. Exclude decision rules with low conformance.
Step 6. Aggregate the remaining decision rules for ease of interpretation.

Figure 3.9 provides a flowchart to illustrate the process, which also shows the interaction with data.

In order to confirm whether the derived input–output relationship conforms to domain knowledge, domain knowledge must be expressed using the same XAI format that was also used to express the input–output relationship.

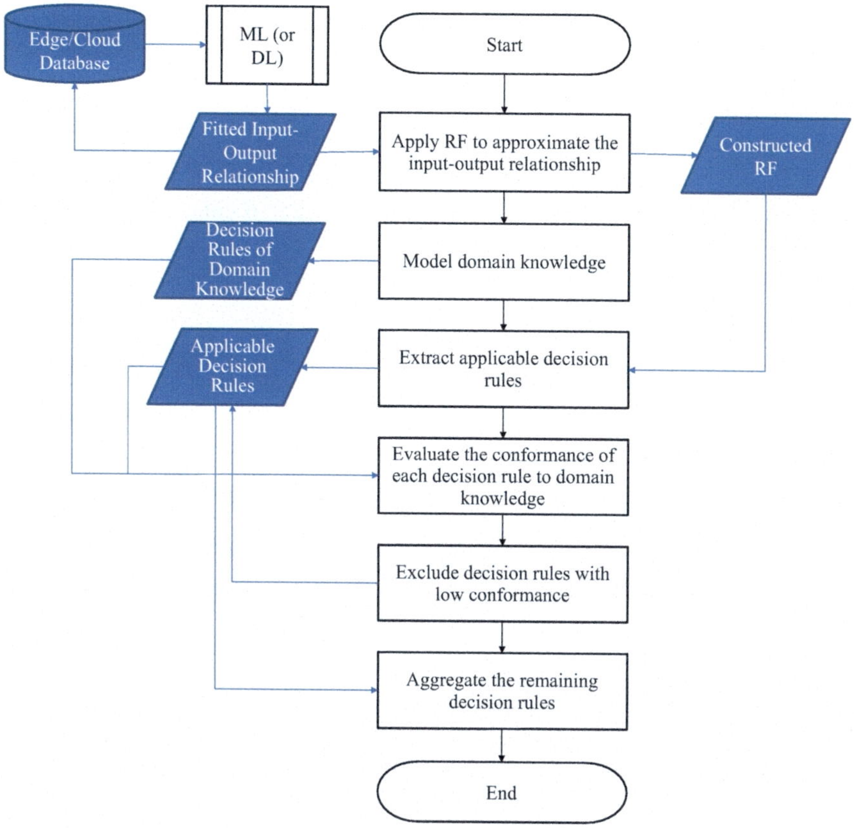

Fig. 3.9 Process of MRFII

In RF, a decision rule has the follow format:

$$\text{“If } x_{j(1)} r_{k(1)} \text{ and} \ldots \text{ and } x_{j(L)} \wedge_{k(L)} r_{k(L)} \text{ then } \hat{o}_j = s_k\text{”} \tag{3.1}$$

where $x_{j(l)}$ is the l-th attribute of object j; $l = 1 \sim L$; $j = 1 \sim n$ (the number of objects). $r_{k(l)}$ is the threshold of $x_{j(l)}$ in the k-th decision rule; $k = 1 \sim K$ (the number of rules). $\wedge_{k(l)} \in \{<, \geq\}$ is the used logical operator. $\hat{o}_j$ is the prediction (detection) result of object j. s_k is the fitted result using rule k.

Chen and Wang (2023) considered the following three domain knowledge:

(1) The most important factors affecting $\hat{o}_j$ include $x_{j(m_1)} \sim x_{j(m_G)}$: : This can be confirmed if a decision tree (or rule) contains all these factors.
(2) $x_{j(m)}$ is more important than $x_{j(p)}$ to $\hat{o}_j$: This can be confirmed if $x_{j(m)}$ appears much earlier (higher) than $x_{j(p)}$ in the decision tree. Since a factor may appear multiple times in a decision tree, only the first occurrence is considered for comparison.

(3) $\hat{o}_j$ increases with $x_{j(m)}$: This can be guaranteed by comparing two decision rules, e.g., decision rules k_1 and k_2, with nearly identical conditions except for the m-th condition:

$$r_{k_1(l)} = r_{k_2(l)} \quad \text{and} \quad \wedge_{k_1(l)} = \wedge_{k_2(l)} \forall l \neq m \tag{3.2}$$

The two decision rules conform to domain knowledge if either of the following conditions is true:

(Condition I)

$$\text{If} \quad \wedge_{k_1(m)} = \wedge_{k_2(m)} \quad \text{then} \quad \begin{cases} s_{k_1} \geq s_{k_2} \; if \; r_{k_1(m)} \geq r_{k_2(m)} \\ s_{k_1} < s_{k_2} \; otherwise \end{cases} \tag{3.3}$$

(Condition II)

$$\text{If} \quad \wedge_{k_1(m)} \neq \wedge_{k_2(m)} \quad \text{then} \quad \wedge_{k_1(m)} r_{k_1(m)} \cap \wedge_{k_2(m)} r_{k_2(m)} \neq \emptyset \tag{3.4}$$

In the third domain knowledge, two decision rules need to be compared, which means.

- Conformities of decision rules are not independent.
- Decision rules applicable to the same object require pairwise comparisons. In addition, decision rules extracted from an ANN (or DNN) using CART or the quick unbiased efficient statistical tree (QUEST) algorithm appear inconsistent (Toth and Eltinge 2011; Bag et al. 2012). Such pairwise comparisons should be made.

In addition, other domain knowledge remains to be considered. However, given so much domain knowledge, experienced engineers in the factory may lack a consensus on some domain knowledge.

Subsequently, the conformity of a decision rule to domain knowledge is evaluated as follows. First, a score is assigned to each domain knowledge. Suppose the score of domain knowledge d is denoted as β_d; $d = 1 \sim D$. Then, the (overall) conformity of decision rule k to domain knowledge can be evaluated by

$$\xi_k = \frac{\sum_{d=1}^{D} I_{kd}}{\sum_{d=1}^{D} \beta_d} \tag{3.5}$$

where

$$I_{kd} = \begin{cases} \beta_d & \text{if rule } k \text{ meets domain knowledge } d \\ 0 & \text{otherwise} \end{cases} \tag{3.6}$$

Decision rules that have too low conformities to domain knowledge are excluded from explaining the estimation result using an ANN (or DNN), as detailed below.

The RF used to explain the estimation process and result using an ANN (or DNN) consists of many decision trees. From each decision tree, a decision rule can be applied to the target object. As a result, multiple rules are applicable simultaneously. To avoid user confusion during interpretation and to make the interpretation conform to domain knowledge (Chen et al. 2024b; Chen 2022a), the extracted decision rules are reorganized as follows:

Step 1. Extract the decision rules applicable to the target object from all decision trees.
Step 2. Evaluate the conformities of these decision rules to domain knowledge.
Step 3. Exclude decision rules that have too low conformities to domain knowledge: The number of remaining decision rules is T.
Step 4. Average the prediction results of the remaining decision rules as $\hat{o}_j$.
Step 5. Set $q = 1$.
Step 6. Select the decision rule which prediction/detection result is closest to $\hat{o}_j$.
Step 7. Let the q-th incremental rule be the decision rule.
Step 8. Replace the prediction/detection result of the decision rule with $\hat{o}_j(current)$.
Step 9. Delete the decision rule.
Step 10. Increase q by 1.
Step 11. If $q > T$, go to Step 17; otherwise, go to Step 12.
Step 12. Find the decision rule which prediction/detection result is closest to $\hat{o}_j$.
Step 13. Subtract $\hat{o}_j(current)$ from the prediction/detection result of the decision rule, and divide the result by T: The result is denoted by $\Delta\hat{o}_j(current)$.
Step 14. Replace the antecedent of the q-th incremental rule with that of the decision rule.
Step 15. Set the result of the q-th incremental rule to "add $\Delta\hat{o}_j(current)$ to $\hat{o}_j$" or "subtract $-\Delta\hat{o}_j(current)$ from $\hat{o}_j$".
Step 16. Return to Step 9.
Step 17. End.

TSMC has extensively deployed ML applications in online edge computing ADC and offline cloud computing ADC, with a particular focus on tool defect prevention at critical manufacturing stages. The function and benefit of ADC are to process more than millions of images every day and improve the sampling rate of image defect detection:

- Inline edge computing ADC, so-called "eagle's view," is embedded in the tool to detect defects while processing and isolating the material. Edge computing is already being used in manufacturing. Most of the existing edge computing applications in manufacturing systems are used for machine status detection (including predictive maintenance) (Chen et al. 2018; Feng et al. 2020), material transportation tracking (Zhao et al. 2020), resource management, process control, production

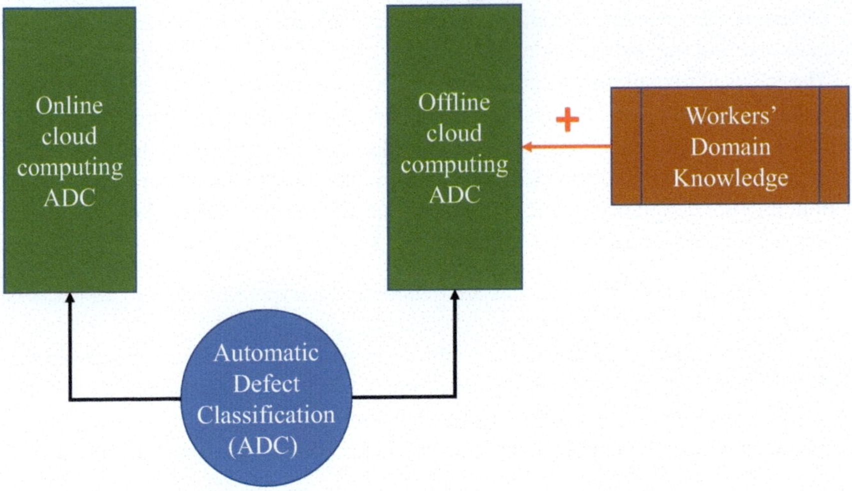

Fig. 3.10 Systematic measures for assisting offline cloud-computed ADC

planning (for machine tools) (Zhang et al. 2021), quality inspection (Wang et al. 2020; Zhu et al. 2020), etc.

- Offline cloud-computed ADC is inserted into process afterward while detecting defects and holding lots to prevent any further loss. The flexibility of ML has been developed and expanded to meet different engineering requirements such as dimensional measurement, defect classification and particle detection to scale up quality management. Making good use of workers' domain knowledge are more helpful to offline cloud-computed ADC (see Fig. 3.10).

3.3.5 Automation

Fab automation reduces dependence on human resources and therefore provides a natural way to avoid cultural differences.

Various automation technologies have been applied to a wafer fab (Chen 2022b; Lee et al. 2023; PEER Group Inc. 2022; Keil 2017). First, a lot of 25 pieces of 300 mm wafers are too heavy for workers to handle. Therefore, the complete automation of material handling is crucial, as automated material handling systems (AMHSs) have been widely used to reduce such high demands on human operators. In this type of wafer fabs, wafers are transported automatically in a front-opening universal pod (FOUP) that can accommodate 25 wafers. Additionally, batch movement is done using AMHSs for inter-bay movement and FOUPs are handed over to rail guided vehicles (RGV) for intra-bay movement. The system is typically 3 to 4 times faster than standard automated guided vehicle (AGV) in-bay systems common in 8-inch wafer fabs (UMC 2024). Other drivers for material handling automation include preventing human handling errors, such as dropped wafers, as well as better tool utilization and faster manufacturing cycle times through fast and reliable material transfer.

In addition, single-wafer processing (SWP) technology, which processes wafers one by one, has been widely introduced in most wafer fabrication processes. To reduce excessive movement tasks between SWP chambers, multiple SWP chambers are integrated in a closed environment together with wafer handling robots (Lee et al. 2023).

In this field, the SEMI Equipment Communication Standard (SECS) and Generic Equipment Model (GEM) protocol standards specify how automated semiconductor manufacturing equipment communicates with a host computer. These standards were developed by the Semiconductor Equipment and Materials International Organization and define interface, messaging and communication functions. Communication can occur via a TCP/IP network or an RS-232 connection (Wagner 2019). This means that the control system of a fully automated equipment can communicate with hosts ranging from local personal computers to cloud servers.

Wafer fabrication equipment, such as chemical delivery systems, cleaning equipment, chemical vapor deposition (CVD) and physical vapor deposition (PVD) equipment, diffusion furnaces, etc., all have embedded sensors and PCs and can therefore be considered cyber-physical systems (CPSs) (Núñez 1990). Status variable identification (SVID) data collected by such sensors can be fed to the DNN for output projection and other purposes (Chen 2022b).

RFID tags attached to wafer carriers can also be used. Forked light barriers attached to a carrier, controlled by an ATMega32 microcontroller, detected the presence of a wafer that has completed fabrication and been ejected by the equipment (Nikitin and Kaski 2021). In this way, the start and finish production times of each job on a specific equipment can be known immediately and accurately. These data are also fed into the DNN for output projection.

Detecting possible anomalies in wafer fabrication equipment based on data collected by embedded sensors can reduce quality problems by 35%, increase production capacity by 10%, and speed up yield learning by 25%, ultimately contributing to more monthly output (Chen 2022b). Such collected data can also be used as inputs to the DNN.

However, there are still human employees in so-called fully automated wafer fabs. They may be gathered in a room remotely monitoring the operation of wafer fabrication equipment and get to the equipment to deal with any problems if arise.

Measures for mitigating the impact of cultural differences are summarized in Fig. 3.11.

3.4 Flexible Long-Term Localization

For a wafer foundry, the pressure and demands from most customers and their governments are unavoidable. However, it is impossible to build so many new wafer fabs around the world at the same time. Instead, the wafer foundry can selectively make

Fig. 3.11 Measures for mitigating the impact of cultural differences

commitments to some customers and their governments to build new wafer fabs in their countries, and arrange for the construction of these new wafer fabs over the long term. To do this, the comparison results of facility locations need to support the development of a flexible long-term localization plan (Chen 2023g; Wang and Chen 2024; Chen and Wang 2024; Wang et al. 2023b, 2023c; Chen et al. 2024c, 2024d) in which multiple facility locations can be selected (see Fig. 3.12).

- **Thresholding**: First, a threshold is established to filter out facility locations with very poor overall performances.
- **Top *k* policy**: Alternatively, a maximum number (k) of new wafer fabs can be determined so that the top k facility locations are selected.
- **Proportionality**: The construction plans at facility locations with better overall performances can be committed earlier, started earlier, and completed earlier. For example, as mentioned earlier, after TSMC's successful experience in building its first new wafer fab in Kumamoto Prefecture, Japan prompted TSMC to immediately launch plans to build two new wafer fabs. For the same reason, the Japanese government was also motivated to provide more financial subsidies and administrative support. As a result, TSMC's fab localization plans in Japan were announced, launched or completed more quickly.

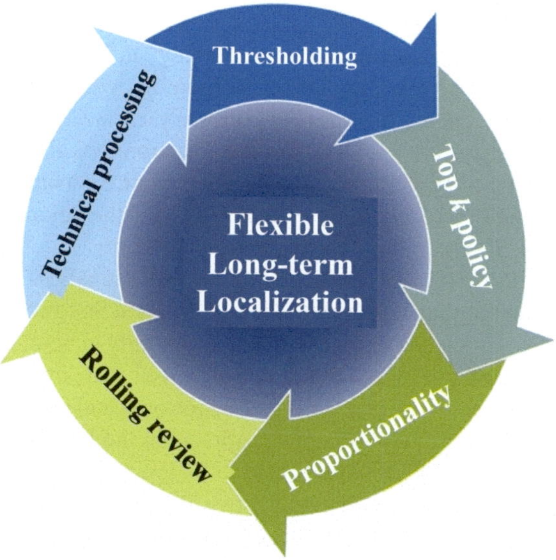

Fig. 3.12 Flexible long-term localization

- **Rolling review**: However, if a facility location become less attractive after some time, the start and completion of the construction plan at the facility location can be postponed.
- **Technical processing**: Finally, if the start (or completion) of the construction plan for a facility location is later than a predetermined time limit, the construction plan is technically cancelled.

3.5 Wafer Technology Node

Wafer technology node has historically referred to many different characteristics of a transistor, including gate length and M1 half-pitch. Lately, due to various marketing and foundry differences, the term no longer represents its previous meanings and has been interpreted differently. The latest technology nodes such as 22, 16, 14 and 10 nm refer to specific generations of wafers manufactured with specific technologies. It does not correspond to any gate length or half pitch. Additionally, transistor size, density, and performance no longer match between foundries. For example, Intel's 10 nm wafer technology node is comparable to the 7 nm wafer technology node in another foundry (WikiChip 2024).

In existing cases of semiconductor supply chain localization, customers often need to establish local capacity capable of manufacturing small/advanced wafer technology nodes. In contrast, the fabrication capacity of mature processes has already been more than sufficient around the world. However, wafer foundries unavoidably keep their top manufacturing technologies domestically to avoid leaking technical secrets.

For example, TSMC's first new wafer fab in Arizona, USA, is expected to start using 5 nm (or 4 nm) process technology in 2025, and TSMC's most advanced wafer fab will also use 2 nm process technology with a monthly capacity of 20,000 wafer starts in the same year. However, the pressure from customers is expected to force this gap to close rapidly. Samsung Electronics is also building a new wafer fab nearby using 4-nm process technology. As a result, Samsung Electronics received up to $6.4 billion in direct funding under CHIPS and the Science Act (James 2021; Electronics et al. 2024).

Shilov (2023) conducted the following analysis of Samsung's localization plan about building new wafer fabs in the United States. Samsung currently makes wafers using its leading fabrication technology only in South Korea, while its S2 wafer fab in Austin, Texas, uses 14 nm low power + (LPP) (and derivatives as low as 11LPP) and older fabrication processes that are too outdated for cutting-edge products (Chen et al. 2020). Samsung has many customers in the US (such as IBM, Nvidia, Qualcomm, Tesla, etc.) that require access to its most advanced nodes and would prefer to use fabs located in the US, which is why Samsung is urging to build a wafer fab there. In addition, the company needs to build a new leading wafer fab in North America to better compete with its rival TSMC, which planned to build wafer fabs in Arizona by 2024.

The analysis result may not be valuable for reference as it often takes more time to push more advanced fabrication processes to a profitable state, i.e., with sufficiently high yields (see Fig. 3.13). New wafer fabs in the U.S. won't solve this problem, but existing fabs in South Korea are more likely to do so. This problem is no exception for other foundries like TSMC.

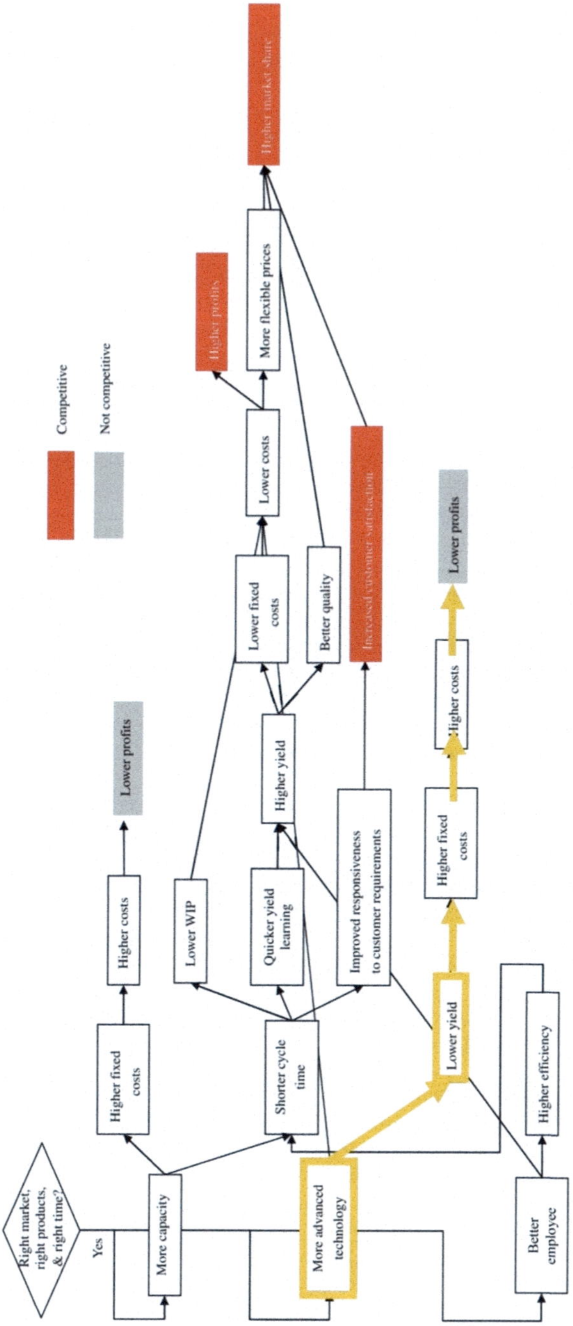

Fig. 3.13 Difficulty in pushing more advanced fabrication processes to a profitable state

References

M. Bag, S. Gauri, S. Chakraborty, Feature-based decision rules for control charts pattern recognition: a comparison between CART and QUEST algorithm. Int. J. Ind. Eng. Comput. **3**(2), 199–210 (2012)

S. Biswas, D. Pamucar, Facility location selection for b-schools in Indian context: a multi-criteria group decision based analysis. Axioms **9**(3), 77 (2020)

Business Weekly, What's wrong with saying "goodbye" to colleagues after get off work? Japanese working in Taiwan: I was shocked. (2024). https://www.businessweekly.com.tw/international/blog/3014886

T. Chen, Y.C. Wang, A modified random forest incremental interpretation method for explaining artificial and deep neural networks in cycle time prediction. Decision Anal. J. **7**, 100226 (2023)

T.C.T. Chen, Y.C. Wang, Evaluating innovative future robotic applications in manufacturing using a fuzzy collaborative intelligence approach. The Int. J. Adv. Manuf. Technol. **130**(11), 6027–6041 (2024)

T. Chen, H.-C. Wu, A partial-consensus and unequal-authority fuzzy collaborative intelligence approach for assessing robotic applications amid the COVID-19 pandemic. Soft. Comput. **27**, 16493–16509 (2023)

B. Chen, J. Wan, A. Celesti, D. Li, H. Abbas, Q. Zhang, Edge computing in IoT-based manufacturing. IEEE Commun. Mag. **56**(9), 103–109 (2018)

T. Chen, C.-W. Lin, Y.-C. Lin, A fuzzy collaborative forecasting approach based on XAI applications for cycle time range estimation. Appl. Soft Comput. **151**, 111122 (2024a)

T.C.T. Chen, H.C. Wu, M.C. Chiu, A deep neural network with modified random forest incremental interpretation approach for diagnosing diabetes in smart healthcare. Appl. Soft Comput. **152**, 111183 (2024d)

T.C.T. Chen, Industry 4.0 for semiconductor manufacturing. in *Production Planning and Control in Semiconductor Manufacturing: Big Data Analytics and Industry 4.0 Applications* (2022), pp. 21–40

T.C.T. Chen, Big data analytics for semiconductor manufacturing. in *Production Planning and Control in Semiconductor Manufacturing: Big Data Analytics and Industry 4.0 Applications* (2022), pp. 1–19

T.C.T. Chen, Job sequencing and scheduling. in *Production Planning and Control in Semiconductor Manufacturing: Big Data Analytics and Industry 4.0 Applications* (2023), pp. 77–100

T.C.T. Chen, Applications of XAI for decision making in the manufacturing domain. in *Explainable Artificial Intelligence (XAI) in Manufacturing: Methodology, Tools, and Applications* (2023), pp. 51–81

T.C.T. Chen, Applications of XAI for forecasting in the manufacturing domain. in *Explainable Artificial Intelligence (XAI) in Manufacturing: Methodology, Tools, and Applications* (2023), pp. 1–11

T. C. T. Chen, Applications of XAI to job sequencing and scheduling in manufacturing. in *Explainable Artificial Intelligence (XAI) in Manufacturing: Methodology, Tools, and Applications* (2023), pp. 83–105

C.-I. Chen, Differences in work culture between Taiwan and the United States (2023). https://translate.google.com/?sl=auto&tl=en&text=%E5%8F%B0%E7%81%A3%E5%92%8C%E7%BE%8E%E5%9C%8B%E7%9A%84%E5%B7%A5%E4%BD%9C%E6%96%87%E5%8C%96%E5%B7%AE%E7%95%B0&op=translate

T.C.T. Chen, Smart technology applications in healthcare before, during, and after the COVID-19 pandemic. in *Sustainable Smart Healthcare: Lessons Learned from the COVID-19 Pandemic* (2023), pp. 19–37

T.C.T. Chen, Explainable artificial intelligence (XAI) in manufacturing. in *Explainable Artificial Intelligence (XAI) in Manufacturing: Methodology, Tools, and Applications* (2023), pp. 1–11

K.Y. Chen, What are the differences in workplace culture between Taiwan and the United States? Why American employees complain that "TSMC is the worst place to work on earth" (2024). https://dq.yam.com/post/16059

T.C.T. Chen, W.A. Chaovalitwongse, M.J. O'grady, K. Honda, Editorial: smart technologies for improving the quality of mobile health care. Health Care Managem. Sci. **23**, 171–172 (2020)

T.C.T. Chen, Y.C. Wang, C.W. Lin, Enhancing the effectiveness of output projection in wafer fabrication using an Industry 4.0 and XAI approach. The Int. J. Adv. Manuf. Technol. 1–13 (2024)

T. Chen, Y.-C. Wang, Y.-C. Lin, Production planning and control of a pharmaceutical supply chain using evolutionary fuzzy approach. J. Eng. Res. 1–10 (2024)

A. Cherrafi, A. Chiarini, A. Belhadi, J. El Baz, A.C. Benabdellah, Digital technologies and circular economy practices: vital enablers to support sustainable and resilient supply chain management in the post-COVID-19 era. The TQM J. **34**(7), 179–202 (2022)

T.Y. Choi, S. Narayanan, D. Novak, J. Olhager, J.B. Sheu, F. Wiengarten, Managing extended supply chains. J. Bus. Logist. **42**(2), 200–206 (2021)

I. Dogan, Analysis of facility location model using Bayesian networks. Expert Syst. Appl. **39**(1), 1092–1104 (2012)

ETtoday, TSMC's factory is very attractive! Japan's Kumamoto Department Store adds a permanent "Taiwan Zone" counter (2024). https://finance.ettoday.net/news/2337039

Y. Feng, Z. Hong, Z. Li, H. Zheng, J. Tan, Integrated intelligent green scheduling of sustainable flexible workshop with edge computing considering uncertain machine state. J. Clean. Prod. **246**, 119070 (2020)

Focus Taiwan, TSMC Kumamoto fab expected to spur Taiwanese investment in Japan (2023). https://focustaiwan.tw/business/202309070018

B. Frieske, S. Stieler, The "semiconductor crisis" as a result of the COVID-19 pandemic and impacts on the automotive industry and its supply chains. World Electric Vehicle J. **13**(10), 189 (2022)

L. James, Why are semiconductor fabs dotting Arizona, the "silicon desert? (2021). https://www.allaboutcircuits.com/news/semiconductor-fabs-dotting-arizona-silicon-desert/

JobMenta, Differences in workplace culture between Taiwan and Japan: Do these 5 points to painlessly adapt to the Japanese workplace (2023). https://jobmenta.yourator.co/blogs/224

C. Kahraman, D. Ruan, I. Doğan, Fuzzy group decision-making for facility location selection. Inf. Sci. **157**, 135–153 (2003)

S. Keil, Design of a cyber-physical production system for semiconductor manufacturing. in *Proceedings of the Hamburg International Conference of Logistics*, vol. 23, pp. 319–340. (2017)

S. Kheybari, M. Kazemi, J. Rezaei, Bioethanol facility location selection using best-worst method. Appl. Energy **242**, 612–623 (2019)

O. Kilincci, S.A. Onal, Fuzzy AHP approach for supplier selection in a washing machine company. Expert Syst. Appl. **38**(8), 9656–9664 (2011)

Y. Koo, S. Kim, 'Follow sourcing' and the transplantation and localization of Korean electronics corporations in northern Vietnam. Area Developm. Policy **7**(4), 380–403 (2022)

T.E. Lee, H.J. Kim, T.S. Yu, Semiconductor manufacturing automation. in *Springer Handbook of Automation* (2023), pp. 841–863

S. Mohseni, K. Baghizadeh, J. Pahl, Evaluating barriers and drivers to sustainable food supply chains. Math. Probl. Eng. **2022**, 4486132 (2022)

S. Nandi, J. Sarkis, A.A. Hervani, M.M. Helms, Redesigning supply chains using blockchain-enabled circular economy and COVID-19 experiences. Sustain. Prod. Consump. **27**, 10–22 (2021)

A. Nikitin, S. Kaski, Decision rule elicitation for domain adaptation. in *26th International Conference on Intelligent User Interfaces* (2021), pp. 244–248

M. Núñez, Decision tree induction using domain knowledge. in *Current Trends in Knowledge Acquisition* (1990), pp. 276–288

PEER Group Inc., SEMI PV2 (2022). https://www.peergroup.com/expertise/resources/semistandards/pv2/

Samsung Electronics, Samsung Electronics to receive up to $6.4 billion in direct funding under the CHIPS and Science Act. (2024). https://news.samsung.com/us/samsung-electronics-to-receive-up-to-6-4-billion-in-direct-funding-under-the-chips-and-science-act/

A. Shilov, Samsung foundry: New $17 billion fab in the USA by late 2023 (2021). https://www.anandtech.com/show/16483/samsung-in-the-usa-a-17-billion-usd-fab-by-late-2023

A. Solucis Santhapparaj, J. Sreenivasan, J., Chong Kuan Loong, Competitive factors of semiconductor industry in Malaysia: The managers' perspectives. Competitiveness Rev.: An Int. Business J. **16**(3/4), 197–211 (2006)

Taipei Times, TSMC Phoenix fab spurs creation of 'Little Taipei' (2022). https://www.taipeitimes.com/News/taiwan/archives/2022/12/08/2003790346

Texas Economic Development Corporation, Texas Instruments to build semiconductor wafer 'fab' plants in Shermax, TX (2021). https://businessintexas.com/news/texas-instruments-to-build-semiconductor-wafer-fab-plants-in-sherman-tx/?utm_source=google&utm_medium=cpc&utm_campaign=PMax&utm_adgroup={adgroup}&creative=&matchtype=&network={networkutm_term=&locphysical=9197734&locinterest=9007609&gad_source=1&gclid=CjwKCAjw7s20BhBFEiwABVIMrRXzD4dhJp8zioELihIWW7C-D321LKWQjcIup4CAGFLF8-8fC4PqzRoCWGwQAvD_BwE

The Asahi Shimbun, Kumamoto enjoys 'bubble' as it prepares for arrival of TSMC (2023). https://www.asahi.com/ajw/articles/14849346

The Economist, TSMC is having more luck building in Japan than in America (2024). https://www.economist.com/business/2024/02/22/tsmc-is-having-more-luck-building-in-japan-than-in-america

The Yomiuri Shimbun, Monitor tour for Taiwanese people in Kumamoto Prefecture, where TSMC has expanded... Planned by department store, includes walking around the castle town and shopping (2024). https://www.yomiuri.co.jp/local/kyushu/news/20240127-OYTNT50170/

A.E. Torkayesh, V. Simic, Stratified hybrid decision model with constrained attributes: recycling facility location for urban healthcare plastic waste. Sustain. Cities Soc. **77**, 103543 (2022)

D. Toth, J.L. Eltinge, Building consistent regression trees from complex sample data. J. Am. Stat. Assoc. **106**(496), 1626–1636 (2011)

TSMC, TSMC celebrates the opening of JASM in Kumamoto, Japan (2024). https://pr.tsmc.com/english/news/3113

TSMC, Automation in packaging fab (2024). https://www.tsmc.com/english/dedicatedFoundry/services/apm_intelligent_packaging_fab/intelligentFab_automation

UMC, UMC is a 12-inch manufacturing leader with several advanced 12-inch fabs in operation (2024). https://www.umc.com/en/Html/fab_information

A. Varas, R. Varadarajan, J. Goodrich, F. Yinug, Government incentives and US competitiveness in semiconductor manufacturing (2020). https://www.eusemiconductors.eu/sites/default/files/uploads/20200916_GovIncentivesUSCompinSCManufacturing.pdf

D. Wagner, How fully automated wafer fabrication equipment improves process control (2019). https://www.modutek.com/how-fully-automated-wafer-fabrication-equipment-improves-process-control/

Y.C. Wang, T. Chen, New XAI tools for selecting suitable 3D printing facilities in ubiquitous manufacturing. Complex and Intell. Syst. **9**(6), 6813–6829 (2023)

Y. Wang, M. Liu, P. Zheng, H. Yang, J. Zou, A smart surface inspection system using faster R-CNN in cloud-edge computing environment. Adv. Eng. Inform. **43**, 101037 (2020)

Y.C. Wang, T.C.T. Chen, Y.C. Lin, 3D printer selection for aircraft component manufacturing using a nonlinear FGM and dependency-considered fuzzy VIKOR approach. Aerospace **10**(7), 591 (2023a)

Y.C. Wang, T. Chen, M.C. Chiu, An explainable deep-learning approach for job cycle time prediction. Decision Anal. J. **6**, 100153 (2023b)

Y.C. Wang, T.C.T. Chen, M.C. Chiu, An improved explainable artificial intelligence tool in healthcare for hospital recommendation. Healthcare Anal. **3**, 100147 (2023c)

Y.C. Wang, T. Chen, Adapted techniques of explainable artificial intelligence for explaining genetic algorithms on the example of job scheduling. Expert Syst. Appl. **237**(A), 121369 (2024)

WikiChip, Technology Node (2024). https://en.wikichip.org/wiki/technology_node

H. Wu, H. Mehrabi, N. Naveed, P. Karagiannidis, Impact of strategic control and supply chain management on recycled plastic additive manufacturing. J. Clean. Prod. **364**, 132511 (2022)

H.-C. Wu, T. Chen, M.-C. Chiu, Assessing the sustainability of smart healthcare applications using a multi-perspective fuzzy comprehensive evaluation approach. Digital Health **9**, 20552076231120390 (2023)

H.-C. Wu, T. Chen, S.Y. Wang, Impact of the COVID-19 pandemic on the revenue of the catering industry: Taiwan as an example. SAGE Open **14**(2), 21582440241241410 (2024)

T. Yunogami, Technology management and competitiveness in the Japanese semiconductor industry. in *Recovering from Success: Innovation and Technology Management in Japan* (2006), pp. 70–86

J. Zhang, C. Deng, P. Zheng, X. Xu, Z. Ma, Development of an edge computing-based cyber-physical machine tool. Robot. Comput.-Integrat. Manuf. **67**, 102042 (2021)

Z. Zhao, P. Lin, L. Shen, M. Zhang, G.Q. Huang, IoT edge computing-enabled collaborative tracking system for manufacturing resources in industrial park. Adv. Eng. Inform. **43**, 101044 (2020)

Z. Zhu, G. Han, G. Jia, L. Shu, Modified densenet for automatic fabric defect detection with edge computing for minimizing latency. IEEE Internet Things J. **7**(10), 9623–9636 (2020)

Chapter 4
Qualitative and Quantitative Analysis of Semiconductor Supply Chain Localization

Abstract This chapter applies some qualitative and quantitative methods to the analysis of semiconductor supply chain localization. First, strength, weakness, opportunity and threat (SWOT) analysis, one of the most well-known qualitative methods for formulating and evaluating strategies, is reviewed. Some examples of SWOT analysis applications in various fields are also given. To demonstrate the application of SWOT analysis in the field of semiconductor supply chain localization, a SWOT analysis is conducted on a major foundry in the world to rationalize the foundry's past choice of building wafer fabs domestically. Subsequently, a SWOT analysis is used again to examine the foundry's localization attempts around the world. A comparative analysis of the localization plans of its two wafer fabs in two countries is also conducted. In the second part of this chapter, two multi-criteria decision making (MCDM) methods, analytical hierarchy process (AHP) and the technique for order preference by similarity to ideal solution (TOPSIS) are combined to compare the suitabilities of several locations for building the localized wafer fab. To further enhance synergies, SWOT analysis and the AHP-TOPSIS method are jointly applied to qualitatively and quantitatively assess the strengths and weaknesses of multiple locations for building the localized wafer fab.

Keywords SWOT analysis · MCDM · Analytical hierarchy process · TOPSIS · Semiconductor supply chain · Localization

4.1 Qualitative Analysis of Semiconductor Supply Chain Localization

4.1.1 SWOT Analysis

SWOT is the combination of strength, weakness, opportunity and threat. It is a common strategic management tool (Samejima et al. 2006) and has been widely used in various fields to demonstrate the strengths and weaknesses of an attempt before delving deeper. For example, **SWOT analysis** can be applied to list and analyze the

T. T. Chen, *Supply Chain Localization in the Semiconductor Industry*,
https://doi.org/10.1007/978-3-031-81280-4_4

strengths and weaknesses, as well as external opportunities and threats, of building a localized wafer fab.

Khan and Al-Ghamdi (2023) applied SWOT analysis to understand the current obstacles and future prospects of the hydrogen-related economy in the Gulf Cooperation Council region. Their study listed the advantages, disadvantages, opportunities and threats related to the hydrogen economy to facilitate subsequent discussions.

Farrokhnia et al. (2024) used SWOT analysis to analyze ChatGPT to understand the advantages and disadvantages of this technology, so that its advantages and weaknesses can be enhanced and overcome.

Cloud manufacturing (CMfg) is a new-generation service-oriented networked manufacturing model that provides centrally managed manufacturing resources, capabilities and services to distributed users (Chen 2022; Chen and Lin 2022a; Ren et al. 2017; Chen and Wang 2021). Chen (2014) discussed the application of CMfg in semiconductor manufacturing plants. The benefits of such applications were divided into five categories: cost savings, efficiency, additional data analysis capabilities, flexibility and closer partnerships. A SWOT analysis was then conducted to guide semiconductor manufacturers in planning their CMfg implementation projects. The simulation of a wafer fabrication plant (wafer fab) was taken as an example. A variety of CMfg services were proposed to assist fab simulation activities through the collaboration of cloud service providers, software vendors, equipment suppliers and wafer fabs. The connection between CMfg applications and wafer fab competitiveness and sustainability was also emphasized.

According to Phadermrod et al. (2019), SWOT analysis has the advantage of being simple and easy to conduct, so it is very popular. However, the disadvantage of SWOT analysis is that the process is too subjective and must be used in conjunction with quantitative methods.

TSMC is the largest wafer foundry, with most of its wafer fabs located in Taiwan (Strong 2023). SWOT analysis can be applied to rationalize TSMC's past choices to build wafer fabs in Taiwan (Jiang 2024). The analysis results are summarized in Fig. 4.1. Obviously, TSMC's past practice of building wafer fabs domestically benefited from government financial and policy support, the cultivation of professional talents, the formation of industrial clusters, the existence of supply chain partners, and the availability of advanced technologies, which enhanced TSMC's competitiveness. But this practice also had weaknesses, such as the dependence on foreign suppliers for raw materials and equipment, energy supply issues and a trend of declining birth rates. Therefore, TSMC should seize opportunities such as investment subsidies in other countries and getting closer to customers and resources. At the same time, TSMC will also face threats such as political instability, excessive foreign investment, and brain drain.

TOWS analysis is a variation of SWOT analysis, created by Weihrich (Weihrich 1982). TOWS and SWOT have the same acronyms with the words in reverse order. TOWS examines a company's external opportunities and threats first, and then compares them to the company's strengths and weaknesses to develop actionable strategies. In short, SWOT analysis is a planning tool, while TOWS analysis is an action tool. If TSMC is taken as an example for conducting a TOWS analysis, TSMC

Strengths	Weaknesses
• Government financial and policy support • Cultivation of professional talents • Industrial clusters • Professional teamwork in the supply chain • Advanced technology	• Reliance on foreign suppliers for raw materials and equipment • Energy supply issues • Trend of declining birthrate
Opportunities	**Threats**
• Investment subsidies in other countries • Geographical location closer to customers and resources	• Political instability • Excessive foreign investment • Brain drain

Fig. 4.1 SWOT analysis for rationalizing TSMC's past choices to build wafer fabs in Taiwan

will also choose to localize its wafer fabs to gain strengths such as government financial and policy support, and to overcome weaknesses such as the dependence on foreign suppliers for raw materials and equipment and energy supply issues. For example, TSMC chose to build a localized wafer fab in Arizona, where the supply of raw materials, equipment and energy is much more abundant than in Taiwan.

4.1.2 SWOT Analysis for Semiconductor Supply Chain Localization

TSMC sees semiconductor supply chain localization as an opportunity to enhance its sustainability and even long-term competitiveness (Frost et al. 2023). The opportunities in its past strategic planning will become strengths through building localized wafer fabs in the United States, Japan and other countries (Chiang 2024):

- **Strengths**: The first strength comes from local governments' policy supports and investment subsidies for the construction of localized wafer fabs in the United States, Japan and other countries. Another benefit comes from tax incentives and exemptions negotiated with local governments, which may be the ultimate deciding factor in choosing a location for building the localized wafer fab (James 2021). Localized wafer fabs are located closer to customers and resources, which is another advantage. In addition, geopolitical and war risks can also be reduced by building new fabrication capacity outside the home country.

- **Weaknesses**: Localized wafer fabs will be affected by cultural differences and suffer from a slower learning progress, which is the first weakness compared with existing wafer fabs. In addition, most locations considered for semiconductor supply chain localization are in developed countries, where wages are relatively high, weakening the profitability of wafer foundries. For the same reason, environmental requirements and occupational laws are generally more stringent in these locations.
- **Opportunities**: The customers of wafer foundries are usually chip designers, and the countries where they are located usually have high-level R&D talents and institutions, which can be an opportunity for wafer foundries to further improve their technical levels. In addition, by moving wafer fabrication capacity closer to customers, there is an opportunity to receive more orders. Furthermore, since the localization of semiconductor supply chains originates from customer requirements, the first to build localized wafer fabs will gain recognition from customers and their governments and receive more subsidies, forming the first-mover advantage.
- **Threats**: In all regions that need localized wafer fabs, huge government subsidies have attracted not only existing wafer foundries, but also local chip designers and semiconductor manufacturers in an attempt to build more fabrication capacity in order to also receive government subsidies. In addition, most of the existing wafer fabs are located domestically, which makes it easier to avoid the leakage of confidential production information and talents from being poached. These will no longer exist after the localization of wafer fabs, forming new threats to the wafer foundry.

The SWAT analysis results are summarized in Fig. 4.2.

Strengths	Weaknesses
• Local governments' policy supports and investment subsidies • Geographical location closer to customers and resources • Less geopolitical and war risks	• Cultural differences • Higher wage levels • Stricter environmental requirements and occupational laws • Slower learning progress
Opportunities	**Threats**
• High-level R&D talents and organizations • More orders from local customers • First-mover advantage	• Competition from local players • Possible leakage of production confidential information • Talents more likely to be poached

Fig. 4.2 SWOT analysis for wafer fab localization

4.1.3 Comparative Analysis

Strengths, weaknesses, opportunities and threats are all relative concepts. Therefore, by conducting a SWOT analysis for a strategic plan against other alternatives can better evaluate the strengths, weaknesses, opportunities, and threats of the plan. For example, TSMC is building new wafer fabs simultaneously in the United States, Japan, and Taiwan. A SWOT analysis of building local wafer fabs in the United States can be more thoroughly analyzed by comparing it to similar plans in Japan:

- It took unexpectedly longer for TSMC to confirm the receipt of US government subsidies (Shilov 2022), showing that the threat and political influence from local manufacturers was greater than expected/planned (Rogers 2021). In comparison, the threat from local players in Japan was almost negligible. The only competitor is United Microelectronics Corporation (UMC), also from Taiwan, which focuses on mature markets. TSMC can get rid of this big threat from UMC by establishing more advanced fabrication processes. In this regard, TSMC enjoys a very obvious advantage over UMC.
- Another threat to the U.S. localization plan of TSMC was the unexpected shortage of construction materials and manpower to build localized wafer fabs and a slower-than-expected recruitment of local employees. In comparison, the construction plan in Japan was surprisingly smooth (TSMC 2024). In addition, compared with the United States, graduates from Japanese universities are more enthusiastic about joining TSMC's localized wafer fabs in Japan. These differences seem to be an advantage for subsequent localized wafer fab construction plans in Japan.
- The concept and methods of just-in-time production (JIT) in Japanese workplaces seem to have a great impact on TSMC workers (Wu et al. 2024; Batth 2021; Chen and Wang 2022, 2023; Wang and Chen 2024). For TSMC, the cultural differences in the United States are greater than those in Japan. Despite efforts by local communities in Japan to eliminate cultural differences, the impact remains huge, deepening the threat to TSMC's localization plans in Japan (Business Weekly 2024). If this problem cannot be properly resolved, it will become more difficult for TSMC to recruit local employees in the future.
- Deep learning (DL) (Chen et al. 2024a; Wang et al. 2023a; Chen and Lin 2022b) relies on graphics processing unit (GPU) acceleration in both the training and inference phases (Buber and Banu 2018). Therefore, generative AI desperately needs GPUs, for which state-of-the-art wafer fabrication technologies are required. Although most GPU chip designers are in the United States, and TSMC has also begun to build localized fabrication capacity there, TSMC currently cannot economically manufacture the required GPU chips in large quantities at its localized wafer fabs in the United States. Only its wafer fabs in Taiwan can do this, which is a short-term weakness for TSMC's localization plan in the United States. In contrast, TSMC's localized wafer fabs in Japan are targeting products for secondary markets and do not have this problem.
- TSMC's localized wafer fabs in Japan use relatively mature wafer fabrication technologies, so the learning progress of these localized wafer fabs will not be

slow (Chen 2023a; Chen and Wang 2020). However, UMC already owns a wafer fab there. Therefore, TSMC's first-mover advantage is not significant.

The comparison results are shown in in Fig. 4.3, where the number of asterisks indicates the relevance.

Strengths	Weaknesses
**** Geographical location closer to customers and resources **** Less geopolitical and war risks *** Local governments' policy supports and investment subsidies	***** Higher wage levels ***** Stricter environmental requirements and occupational laws **** Cultural differences **** Unable to meet the needs of generative AI *** Slower learning progress
Opportunities	**Threats**
**** High-level R&D talents and organizations *** More orders from local customers *** First-mover advantage	***** Competition from local players ***** Possible leakage of production confidential information ***** Talent more likely to be poached

(Localization Plan of TSMC in the U.S.)

Strengths	Weaknesses
***** Local governments' policy supports and investment subsidies ***** Less geopolitical and war risks **** Geographical location closer to customers and resources	**** Higher wage levels *** Cultural differences *** Stricter environmental requirements and occupational laws ** Slower learning progress
Opportunities	**Threats**
***** High-level R&D talents and organizations **** More orders from local customers *** First-mover advantage (compared to foundries other than UMC)	*** Possible leakage of production confidential information ** Competition from local players ** Talent more likely to be poached

(Localization Plan of TSMC in Japan)

Fig. 4.3 Comparison results

4.2 Quantitative Analysis of Semiconductor Supply Chain Localization

Choosing the most suitable location to build a localized wafer fab requires consideration of many factors, leading to a multi-criteria decision-making (MCDM) problem. In order to solve this problem, Jiang (2024) proposed the analytical hierarchy process (AHP) and technique for order preference by similarity to ideal solution (TOPSIS) approach. AHP and TOPSIS are both popular MCDM methods (Chen et al. 2023a, 2021; Wu et al. 2020; Chen 2021a; Nazar et al. 2019).

4.2.1 *The AHP-TOPSIS Approach*

The localization of wafer fabs is believed to help the sustainable operations of wafer foundries, which includes three major aspects: social, environmental and economic aspects (Chen 2014). In these aspects, six criteria for evaluating the suitability of a location for building a localized wafer fab can be listed: energy costs, water costs, natural disaster risks, human resource costs, technical talent adequacy and political stability. The structure and quantification of these criteria are shown in Table 4.1. Energy costs, water costs, natural disaster risks, human resource costs and political stability are the-lower-the-better (LTB) measures, while technical talent adequacy is the higher-the-better (HTB) measure.

Locations in seven countries were evaluated by Jiang (2024): the United States, Japan, Germany, France, the United Kingdom, Canada and South Korea. The characteristics of these countries for building localized wafer fabs were compared in Table 4.2 (Jiang 2024). The decision-making hierarchy for this semiconductor supply chain localization problem is shown in Fig. 4.4. The AHP-TOPSIS approach can be applied to solve this problem.

At first, the decision maker compares the relative priorities of criteria for evaluating the suitability of a location for building a localized wafer fab, thereby constructing a judgment (or pairwise comparison) matrix $\mathbf{A} = \{a_{ij}\}$. a_{ij} is the relative priority of criterion i over criterion j; $i, j = [1, n]$. In practical applications, a_{ij} is usually chosen from 1 to 9 and represented by a linguistic term (see Table 4.3).

w_i is the priority of criterion i. To derive the value of w_i, the sum of square deviations is minimized according to Satty's method (Satty 1980)

$$\text{Min } SSD = \sum_{i=1}^{n} \sum_{j \neq i} \left(\frac{w_i}{w_j} - a_{ij}\right)^2 \tag{4.1}$$

The optimal solution can be obtained by performing an eigen analysis:

$$det(\mathbf{A} - \lambda \mathbf{I}) = 0 \tag{4.2}$$

Table 4.1 Structure and quantification of the criteria

Aspect	Code	Criterion	Quantification
Economic	S1	Energy costs	Energy cost is related to the average energy price in the region, calculated based on the industrial electricity price of each country in 2021
	S2	Water costs	Water cost is the average unit price of 100 degrees of water in 2021
Environmental	S3	Natural disaster risks	Natural disaster risk is the probability of natural disasters occurring in the region. The data is taken from the statistical data in 2016, calculated by the Institute for Environment and Human Security (EHS) of the United Nations University (UNU) and Bündnis Entwicklung hilft (BEH) (Wikipedia 2024a)
Social	S4	Human resource costs	The average salary of labor in the country, calculated based on the minimum hourly wage in 2021
	S5	Technical talent adequacy	The number of the country's universities in Quacquarelli Symonds (QS) 2023 world's top 500 universities in electronic or electrical engineering
	S6	Political stability	Measured in terms of the 2021 Fragile State Index The Fund for Peace (2024)

$$(\mathbf{A} - \lambda \mathbf{I})\mathbf{x} = 0 \tag{4.3}$$

$$\mathbf{w} = N(\mathbf{x}) \tag{4.4}$$

where λ and $\mathbf{x}$ are the eigenvalue and eigenvector of $\mathbf{A}$, respectively; *det*() is the determinant function; *N*() is the normalisation function. However, the results of pairwise comparisons need to be consistent:

$$CR = \frac{\lambda_{\max} - n}{n - 1} / RI \leq 0.1 \tag{4.5}$$

where *CR* is the consistency ratio (Satty 1980) and *RI* is the random consistency index (Satty 1980). Otherwise, pairwise comparisons need to be redone.

Example 4.1 The judgment matrix constructed by a decision maker for the semiconductor supply chain localization problem is

Table 4.2 Characteristics of various countries for building localized wafer fabs

Country	Advantages	Disadvantages
U.S	• National policies and subsidies • 60% of customers are located there • Many talents	• High labor costs • Political factors have a great influence (the US-China trade war)
Japan	• Proximity to key customers • National policies and subsidies • Abundant energy resources	• Shortage of human resources • Insufficient industrial land
Germany	• Huge market	• Insufficient talents • High energy costs
France	• The semiconductor industry is complete and mature • The output value is second only to Germany in Europe	• Labor costs are high • National policies support its local fabs and are detrimental to Asian fabs • The market is small
U.K	• Increased wafer demand	• Incomplete supply chains • The semiconductor industry is small • Not technologically advanced
Canada	• Proximity to key customers • Abundant energy resources • Tariff preferences • Many talents	• Industrial development is not yet mature • Insufficient supporting government policies and measures
South Korea	• National policies and subsidies	• Lack of high-level technical talents

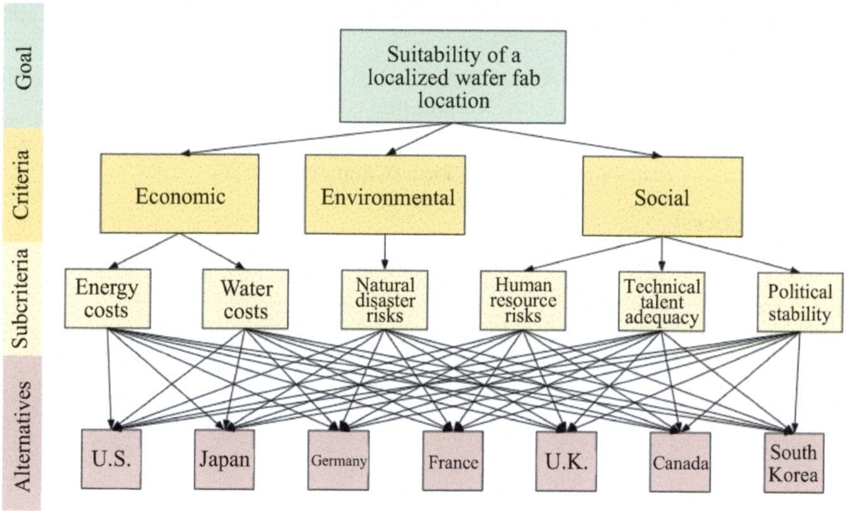

Fig. 4.4 Decision-making hierarchy

Table 4.3 Linguistic terms for representing the relative priorities

a_{ij}	Linguistic term
1	As equal as
2	As equal as or weakly more important than
3	Weakly more important than
4	Weakly or strongly more important than
5	Strongly more important than
6	Strongly or very strongly more important than
7	Very strongly more important than
8	Very strongly or absolutely more important than
9	Absolutely more important than

$$\mathbf{A} = \begin{bmatrix} 1 & 1 & 5 & 1/3 & 3 & 3 \\ 1 & 1 & 5 & 1/3 & 3 & 3 \\ 1/5 & 1/5 & 1 & 1/9 & 1/3 & 1/3 \\ 3 & 3 & 9 & 1 & 5 & 5 \\ 1/3 & 1/3 & 3 & 1/5 & 1 & 1 \\ 1/3 & 1/3 & 3 & 1/5 & 1 & 1 \end{bmatrix}$$

$CR(\mathbf{A}) = 0.021 \leq 0.1$, implying that the pairwise comparison results are consistent and the evaluation results will be reliable. The priorities of criteria for evaluating the suitability of a location for building a localized wafer fab are derived from **A** and listed in Table 4.4.

Subsequently, FTOPSIS method is applied to evaluate and compare the overall performances of possible locations for building a localized wafer fab.

First, the performances of various locations in optimizing each criterion are normalized using distributive normalization (Chen et al. 2021):

(HTB Measures)

$$\rho_{qi} = \frac{p_{qi}}{\sqrt{\sum_{\phi=1}^{Q} p_{\phi i}^2}} = \frac{1}{\sqrt{1 + \sum_{\phi \neq q} (\frac{p_{\phi i}}{p_{qi}})^2}} \tag{4.6}$$

Table 4.4 Priorities of criteria for evaluating the suitability of a location for building a localized wafer fab

w_1	w_2	w_3	w_4	w_5	w_6
0.191	0.191	0.034	0.432	0.076	0.076

(LTB Measures)

$$\rho_{qi} = 1 - \frac{p_{qi}}{\sqrt{\sum_{\phi=1}^{Q} p_{\phi i}^2}} = 1 - \frac{1}{\sqrt{1 + \sum_{\phi \neq q} (\frac{p_{\phi i}}{p_{qi}})^2}} \tag{4.7}$$

where p_{qi} is the performance of the q-th location in optimizing the i-th criterion; ρ_{qi} is the normalized performance.

Then, prioritized scores (s_{qi}) are calculated based on the derived priorities (Wikipedia 2024a):

$$s_{qi} = w_i \rho_{qi} \tag{4.8}$$

The ideal (zenith) point and the anti-ideal (nadir) point are specified respectively as (Wikipedia 2024a)

$$\mathbf{\Lambda}^+ = \{\Lambda_i^+\} = \{\max_q s_{qi}\} \tag{4.9}$$

$$\mathbf{\Lambda}^- = \{\Lambda_i^-\} = \{\min_q s_{qi}\} \tag{4.10}$$

The distances from location q to the two reference points are measured respectively as (Wikipedia 2024a)

$$d_q^+ = \sqrt{\sum_{i=1}^{n} (\Lambda_i^+ - s_{qi})^2} \tag{4.11}$$

$$d_q^- = \sqrt{\sum_{i=1}^{n} (s_{qi} - \Lambda_i^-)^2} \tag{4.12}$$

Finally, the closeness (i.e., overall performance) of location q for building the localized wafer fab is evaluated as (Wikipedia 2024a)

$$C_q = \frac{d_q^-}{d_q^+ + d_q^-} \tag{4.13}$$

Example 4.2 The performances of the seven alternatives in Fig. 4.4 are summarized in Table 4.5 (Jiang 2024). Their normalization results are presented in Table 4.6.

After multiplying the priorities of criteria to the normalized performances, the prioritized scores are derived and shown in Table 4.7.

For this case, the two references points are specified and presented in Table 4.8.

Table 4.5 Performances of the seven alternatives.

Alternative	Symbol	Energy costs (NTD$)	Water costs (NTD$)	Natural disaster risks	Human resource costs (USD$)	Technical talent adequacy	Political stability
U.S	A1	2.04	64.1	3.76	7.25	81	44.6
Japan	A2	4.11	32.1	12.99	8.47	16	32.2
Germany	A3	5.21	58.3	2.95	11.24	15	24.8
France	A4	3.81	62.4	2.62	12.12	17	32.5
U.K	A5	5.27	75.0	3.54	12.25	35	41.5
Canada	A6	2.60	96.9	3.01	11.44	22	21.7
South Korea	A7	2.68	15.8	4.59	7.62	18	32.5

Table 4.6 Normalized performances of the seven alternatives

q	ρ_{q1}	ρ_{q2}	ρ_{q3}	ρ_{q4}	ρ_{q5}	ρ_{q6}
1	0.800	0.615	0.758	0.733	0.837	0.500
2	0.598	0.807	0.163	0.688	0.165	0.639
3	0.490	0.650	0.810	0.586	0.155	0.722
4	0.627	0.625	0.831	0.553	0.176	0.635
5	0.484	0.549	0.772	0.549	0.362	0.534
6	0.746	0.418	0.806	0.578	0.227	0.757
7	0.738	0.905	0.704	0.719	0.186	0.635

Table 4.7 Prioritized scores of the seven alternatives

q	s_{q1}	s_{q2}	s_{q3}	s_{q4}	s_{q5}	s_{q6}
1	0.153	0.117	0.026	0.317	0.064	0.038
2	0.114	0.154	0.006	0.297	0.013	0.049
3	0.094	0.124	0.028	0.253	0.012	0.055
4	0.120	0.119	0.028	0.239	0.013	0.048
5	0.093	0.105	0.026	0.237	0.027	0.041
6	0.142	0.080	0.027	0.250	0.017	0.057
7	0.141	0.173	0.024	0.311	0.014	0.048

Table 4.8 References points for this case

	$\Lambda_1^{+/-}$	$\Lambda_2^{+/-}$	$\Lambda_3^{+/-}$	$\Lambda_4^{+/-}$	$\Lambda_5^{+/-}$	$\Lambda_6^{+/-}$
$\mathbf{\Lambda}^+$	0.153	0.173	0.028	0.317	0.064	0.057
$\mathbf{\Lambda}^-$	0.093	0.080	0.006	0.237	0.012	0.038

Table 4.9 Evaluation results using TOPSIS

q	d_q^+	d_q^+	C_q	Rank
1	0.059	0.120	0.672	2
2	0.074	0.099	0.572	3
3	0.112	0.055	0.327	4
4	0.112	0.054	0.326	5
5	0.127	0.036	0.222	7
6	0.124	0.060	0.324	6
7	0.052	0.130	0.713	1

The closeness (i.e., overall performance) of each location for building the localized wafer fab is evaluated. The evaluation results are summarized in Table 4.9.

The seven locations are ranked based on their closeness values. A location is more suitable for building the localized wafer fab if its closeness is higher. In this case, the most suitable location for building the localized wafer fab is location #7, i.e., South Korea, while the least suitable location for building the localized wafer fab is location #5, i.e., U.K.

4.2.2 *Visualizing the Evaluation Results*

To visualize the evaluation results, an explainable artificial intelligence (XAI) tool, the bidirectional scatterplot (Wang and Chen 2023), can be applied (see Fig. 4.5):

- The ideal solution is at the top, while the anti-ideal solution is at the bottom.
- The ideal solution is white, while the anti-ideal solution is black.

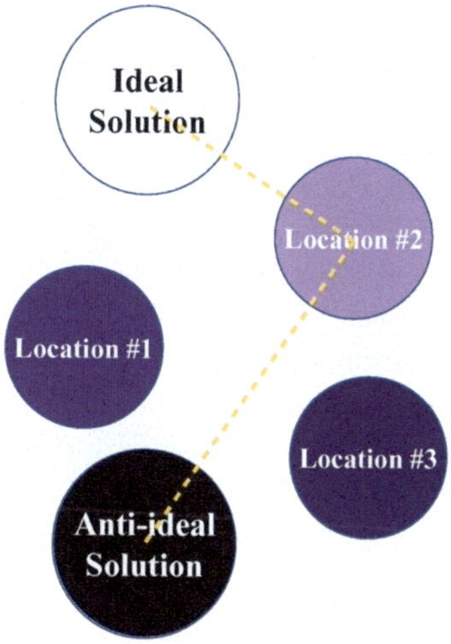

Fig. 4.5 Example of bidirectional scatterplot

Fig. 4.6 Bidirectional scatterplot result

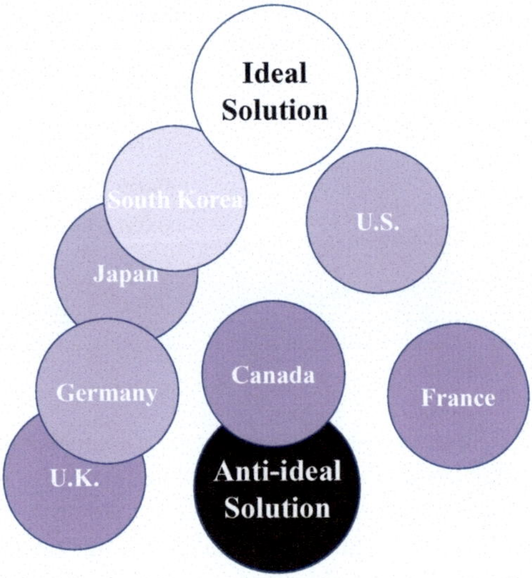

- All locations for building the localized wafer fab are placed anywhere in the bidirectional scatter plot if the following requirement is met. The distances from each location to the two reference points are measured using the TOPSIS method, as illustrated in this figure.
- The closer a location for building the localized wafer fab is to the ideal solution, the whiter (lighter) its color is. Conversely, a location becomes blacker if it is closer to the anti-ideal solution.

Example 4.3 For the previous case, a bidirectional scatterplot is drawn in Fig. 4.6 to visualize the evaluation results to highlight the differences between the overall performances of various locations for building the localized wafer fab.

Shapley additive explanations (SHAP) a game theory method for interpreting the output of any machine learning (ML) or deep learning (DL) model. It uses the classic Shapley value from game theory and its related extensions to connect optimal credit allocation with local explanations (Chen 2024). SHAP analysis is very popular in assessing the importance of inputs to outputs of ML or DL applications (Lundberg 2018; Chen et al. 2024b; Nguyen et al. 2021; Chen 2023b, 2023c; Roshan and Zafar 2022), and can also be applied to MCDM methods if the decision-making result is not obvious from the option attributes (Pillai 2024; Chen 2023d).

Take the first option, the United States, as an example. The importance of each criterion to this option is evaluated based on its SHAP value, for which a waterfall plot is drawn in Fig. 4.7. In addition, for the same option, the SHAP values of all criteria are summarized by a force plot, as shown in Fig. 4.8. Furthermore, the SHAP analysis results of all option can be summarized by the a beeswarm plot in Fig. 4.9. The relationship between the value and SHAP value of a criterion can be visualized using a scatter plot. Take the first criterion, energy costs, as an example. The result is shown in Fig. 4.10.

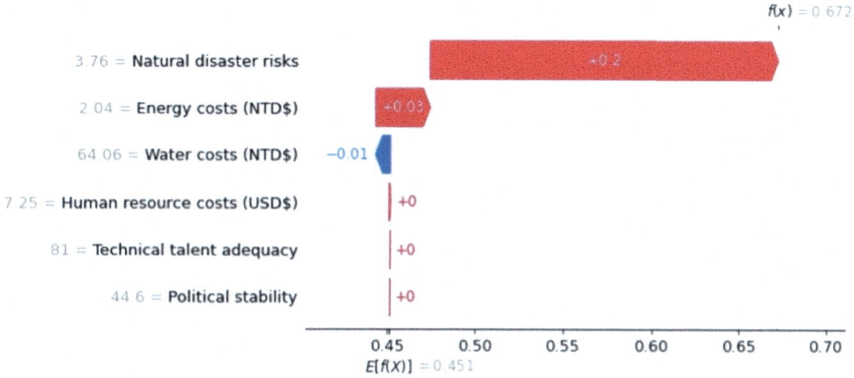

Fig. 4.7 Waterfall plot

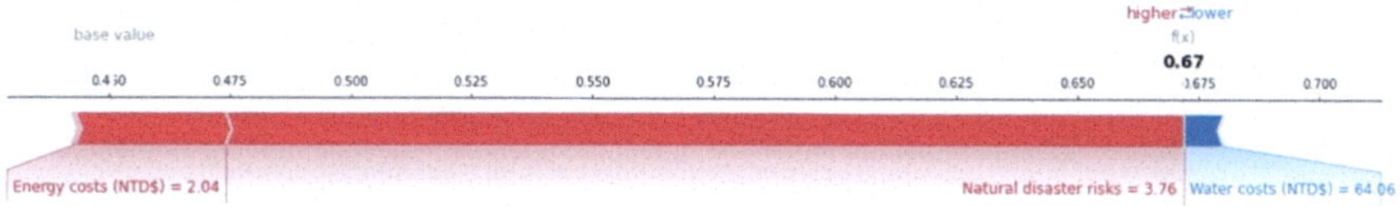

Fig. 4.8 Force plot for the same option

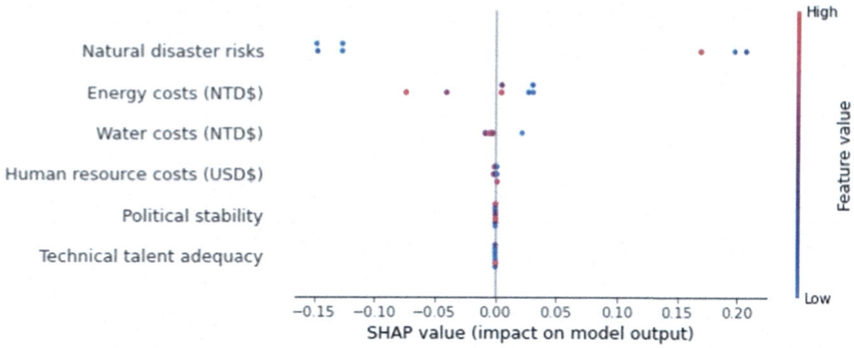

Fig. 4.9 Beeswarm plot for all options

4.2.3 *Implication for the Practice*

The evaluation results have the following implications for the practice:

- In this case, the most suitable location for building the localized wafer fab is in South Korea. However, Samsung in this country is one of TSMC's biggest competitors. This fact was not considered in the decision-making process because relevant factors were not taken into account.

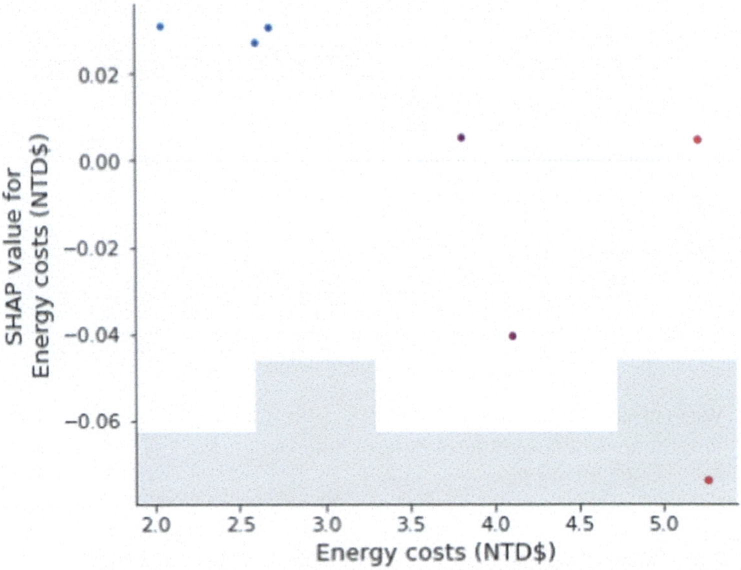

Fig. 4.10 Scatter plot

- TSMC chose to build localized wafer fabs in U.S. and Japan, which rank 2nd and 3rd according to the evaluation results.
- TSMC needs to examine why the location in South Korea is better than those in U.S. and Japan for building the localized wafer fab. Based on Table 4.7, it is clear that part of the reason is the very low cost of water in South Korea. In addition, South Korea's human resource costs and energy costs are also relatively low.

4.2.4 *Accounting for Uncertainty Using Fuzzy Logic*

Semiconductor supply chain localization is a long-term task. Therefore, there will be many uncertainties during the fulfillment of this task, which may weaken the optimality of the decision-making result. To address this issue, methods combined with fuzzy logic can be applied, such as the fuzzy analytic hierarchy process (FAHP)-fuzzy technique for order preference by similarity to ideal solution (FTOPSIS) approach (Chen et al. 2023a).

The FAHP-FTOPSIS approach comprises the following steps:

Step 1. List factors to be considered in choosing a suitable location for building the localized wafer fab in a global semiconductor supply chain.
Step 2. Establish criteria corresponding to these factors.
Step 3. Compare the relative priorities of criteria in pairs.
Step 4. Derive the fuzzy priorities of criteria.

Step 5. Collect (or evaluate) the performances of possible locations for building the localized wafer fab in optimization each criterion.
Step 6. Derive and compare the overall performances of possible locations for building the localized wafer fab using FTOPSIS.
Step 7. Recommend the top performing locations.

Figure 4.11 provides a flowchart to illustrate the procedure for implementing the FAHP-FTOPSIS approach.

In the FAHP-FTOPSIS approach, the decision maker constructs a fuzzily consistent judgment matrix $\tilde{\mathbf{A}} = \{\tilde{a}_{ij}\}$ instead. $\tilde{a}_{ij}$ is mapped to a triangular fuzzy number (TFN) within (Samejima et al. 2006; Phadermrod et al. 2019) (see Table 4.10) (Dong et al. 2021; Chen 2021b).

Definition 4.1 A TFN $\tilde{A} = (A_1, A_2, A_3)$ is a fuzzy number with the following membership function:

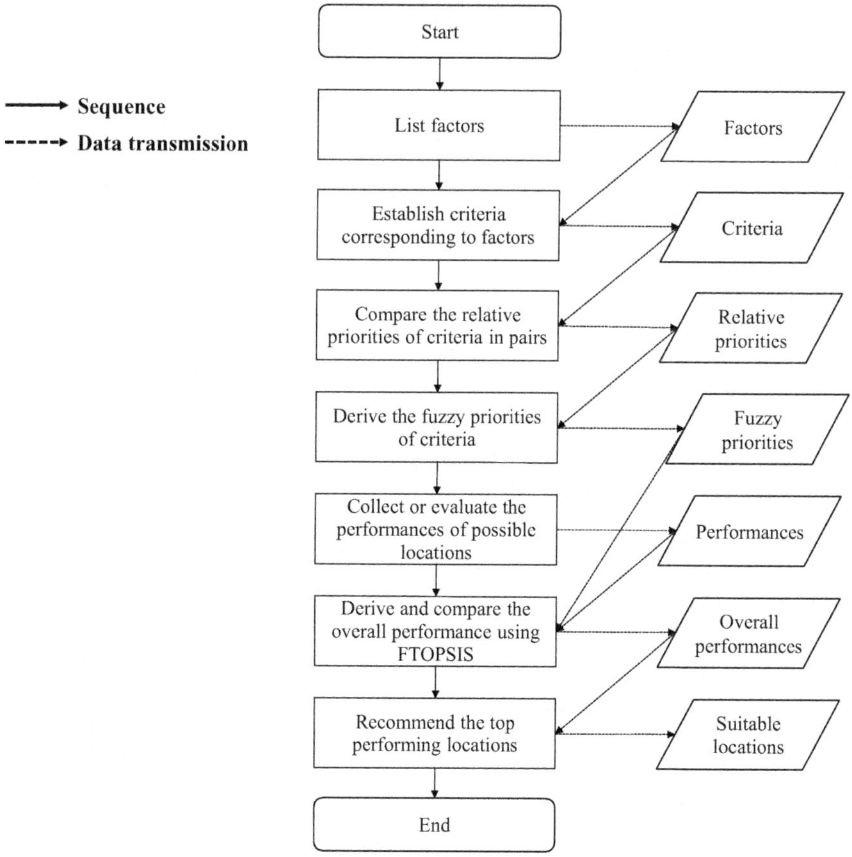

Fig. 4.11 Procedure for implementing the FAHP-FTOPSIS approach

Table 4.10 Linguistic terms for representing the fuzzy relative priorities

$\tilde{a}_{ij}$	Linguistic term
(1, 1, 3)	As equal as
(1, 2, 4)	As equal as or weakly more important than
(1, 3, 5)	Weakly more important than
(2, 4, 6)	Weakly or strongly more important than
(3, 5, 7)	Strongly more important than
(4, 6, 8)	Strongly or very strongly more important than
(5, 7, 9)	Very strongly more important than
(6, 8, 9)	Very strongly or absolutely more important than
(7, 9, 9)	Absolutely more important than

$$\mu_{\tilde{A}}(x) = \begin{cases} \frac{x-A_1}{A_2-A_1} \; if & A_1 \le x < A_2 \\ \frac{A_3-x}{A_3-A_2} \; if & A_2 \le x < A_3 \\ 0 \quad otherwise & \end{cases} \tag{4.14}$$

Some arithmetic operations on TFNs are summarized as follows:

- Fuzzy addition:

$$\tilde{A}(+)\tilde{B} = (A_1 + B_1, A_2 + B_2, A_3 + B_3) \tag{4.15}$$

- Fuzzy subtraction:

$$\tilde{A}(-)\tilde{B} = (A_1 - B_3, A_2 - B_2, A_3 - B_1) \tag{4.16}$$

- Fuzzy multiplication:

$$\tilde{A}(\times)\tilde{B} \cong (A_1B_1, A_2B_2, A_3B_3) \tag{4.17}$$

if $A_1, B_1 \ge 0$.

- Fuzzy division:

$$\tilde{A}(/)\tilde{B} \cong (A_1/B_3, A_2/B_2, A_3/B_1) \tag{4.18}$$

if $A_1 \ge 0, B_1 > 0$.

To derive the value of the fuzzy priority of criterion i, $\tilde{w}_i$, the fuzzy sum of square deviations is minimized:

$$\text{Min } FSSD = \sum_{i=1}^{n} \sum_{j \ne i} (\frac{\tilde{w}_i}{\tilde{w}_j}(-)\tilde{a}_{ij})^2 \tag{4.19}$$

The optimal solution can be obtained using a fuzzy eigen analysis as

$$\det(\tilde{\mathbf{A}}(-)\tilde{\lambda}\mathbf{I}) = 0 \tag{4.20}$$

$$(\tilde{\mathbf{A}}(-)\tilde{\lambda}\mathbf{I})(\times)\tilde{\mathbf{x}} = 0 \tag{4.21}$$

$$\tilde{\mathbf{w}} = [\tilde{w}_i] = \frac{\tilde{\mathbf{x}}}{|\tilde{\mathbf{x}}|} \tag{4.22}$$

where $\tilde{\lambda}$ and $\tilde{\mathbf{x}}$ denote the fuzzy eigenvalue and eigenvector of $\tilde{\mathbf{A}}$, respectively. $(-)$ and $(\times)$ are fuzzy subtraction and multiplication operations, respectively. However, involving fuzzy multiplication operations makes (4.20) and (4.21) intractable (Chen et al. 2023b; Wang and Chen 2022; Chen and Chiu 2022). To address this issue, Chen et al. (2023a) proposed the selectively calibrated derivation technique (SCDT) as follows:

(SCDT)

Step 1. Approximate the value of the fuzzy priority of criterion *i* using fuzzy geometric mean (FGM) as (Buckley 1985; Chen 2023e, 2021c; Huynh et al. 2018):

$$w_{i1}(\text{FGM}) = \frac{1}{1 + \sum_{m \neq i} \frac{\sqrt[n]{\prod_{j=1}^{n} a_{mj3}}}{\sqrt[n]{\prod_{j=1}^{n} a_{ij1}}}} \tag{4.23}$$

$$w_{i2}(\text{FGM}) \cong \frac{1}{1 + \sum_{m \neq i} \frac{\sqrt[n]{\prod_{j=1}^{n} a_{mj2}}}{\sqrt[n]{\prod_{j=1}^{n} a_{ij2}}}} \tag{4.24}$$

$$w_{i3}(\text{FGM}) \cong \frac{1}{1 + \sum_{m \neq i} \frac{\sqrt[n]{\prod_{j=1}^{n} a_{mj1}}}{\sqrt[n]{\prod_{j=1}^{n} a_{ij3}}}} \tag{4.25}$$

Step 2. Approximate the value of the fuzzy priority of criterion *i* using fuzzy inverse of column sum (FICSM) as (Ahmed and Kilic 2019):

$$w_{i1}(\text{FICSM}) = \frac{1}{1 + \sum_{j \neq i} \frac{\sum_{m=1}^{n} a_{mi3}}{\sum_{m=1}^{n} a_{mj1}}} \tag{4.26}$$

$$w_{i2}(\text{FICSM}) = \frac{1}{1 + \sum_{j \neq i} \frac{\sum_{m=1}^{n} a_{mi2}}{\sum_{m=1}^{n} a_{mj2}}} \tag{4.27}$$

$$w_{i3}(\text{FICSM}) = \frac{1}{1 + \sum_{j \neq i} \frac{\sum_{m=1}^{n} a_{mi1}}{\sum_{m=1}^{n} a_{mj3}}} \tag{4.28}$$

Step 3. Derive the crisp priority of criterion i as $w_i(\text{CRISP})$.

Step 4. If $\sum_{i=1}^{n} |w_{i2}(\text{FICSM}) - w_i(\text{CRISP})| \leq \sum_{i=1}^{n} |w_{i2}(\text{FGM}) - w_i(\text{CRISP})|$, approximate the fuzzy priority of criterion i as

$$w_{i1}(\text{SCDT}) = w_{i1}(\text{FICSM}) \cdot \frac{w_i(\text{CRISP})}{w_{i2}(\text{FICSM})} \tag{4.29}$$

$$w_{i2}(\text{SCDT}) = w_{i2}(\text{FICSM}) \cdot \frac{w_i(\text{CRISP})}{w_{i2}(\text{FICSM})} \tag{4.30}$$

$$w_{i3}(\text{SCDT}) = w_{i3}(\text{FICSM}) \cdot \frac{w_i(\text{CRISP})}{w_{i2}(\text{FICSM})} \tag{4.31}$$

otherwise,

$$w_{i1}(\text{SCDT}) = w_{i1}(\text{FGM}) \cdot \frac{w_i(\text{CRISP})}{w_{i2}(\text{FGM})} \tag{4.32}$$

$$w_{i2}(\text{SCDT}) = w_{i2}(\text{FGM}) \cdot \frac{w_i(\text{CRISP})}{w_{i2}(\text{FGM})} \tag{4.33}$$

$$w_{i3}(\text{SCDT}) = w_{i3}(\text{FGM}) \cdot \frac{w_i(\text{CRISP})}{w_{i2}(\text{FGM})} \tag{4.34}$$

Subsequently, FTOPSIS method is applied to evaluate and compare the overall performances of possible locations for building the localized wafer fab. First, the performance of location q in optimizing each criterion is normalized using fuzzy distributive normalization as

$$\tilde{\rho}_{qi} = \frac{\tilde{p}_{qi}}{\sqrt{\sum_{\phi=1}^{Q} \tilde{p}_{\phi i}^{2}}} = \frac{1}{\sqrt{1 + \sum_{\phi \neq q} (\frac{\tilde{p}_{\phi i}}{\tilde{p}_{qi}})^2}} \tag{4.35}$$

where $\tilde{p}_{qi}$ is the performance of the q-th location in optimizing the i-th criterion; $\tilde{\rho}_{qi}$ is the normalized performance. Obviously,

$$\rho_{qi1} = \frac{1}{\sqrt{1 + \sum_{j \neq i} (\frac{p_{qj3}}{p_{qi1}})^2}} \tag{4.36}$$

$$\rho_{qi2} = \frac{1}{\sqrt{1 + \sum_{j \neq i} (\frac{p_{qj2}}{p_{qi2}})^2}} \tag{4.37}$$

$$\rho_{qi3} = \frac{1}{\sqrt{1 + \sum_{j \neq i} (\frac{p_{qj1}}{p_{qi3}})^2}} \tag{4.38}$$

Fuzzy prioritized scores are then calculated based on the fuzzy priorities derived using SCDT:

$$\tilde{s}_{qi} = \tilde{w}_i(\times)\tilde{\rho}_{qi} \tag{4.39}$$

Equivalently,

$$s_{qi1} = w_{i1}\rho_{qi1} \tag{4.40}$$

$$s_{qi2} = w_{i2}\rho_{qi2} \tag{4.41}$$

$$s_{qi3} = w_{i3}\rho_{qi3} \tag{4.42}$$

Fuzzy ideal (zenith) point and fuzzy anti-ideal (nadir) point are then specified respectively as

$$\tilde{\Lambda}^+ = \{\tilde{\Lambda}_i^+\} = \{\max_q \tilde{s}_{qi}\} \tag{4.43}$$

$$\tilde{\Lambda}^- = \{\tilde{\Lambda}_i^-\} = \{\min_q \tilde{s}_{qi}\} \tag{4.44}$$

The fuzzy distances from location q to the two reference points are measured respectively as

$$\begin{aligned} \tilde{d}_q^+ &= \sqrt{\sum_{i=1}^{n} (\tilde{\Lambda}_i^+(-)\tilde{s}_{qi})^\eta} \\ &\cong (\sqrt{\sum_{i=1}^{n} \max(0,\ \Lambda_{i1}^+ - s_{qi3})^\eta},\ \sqrt{\sum_{i=1}^{n} (\Lambda_{i2}^+ - s_{qi2})^\eta},\ \sqrt{\sum_{i=1}^{n} (\Lambda_{i3}^+ - s_{qi1})^\eta}) \end{aligned} \tag{4.45}$$

$$\tilde{d}_q^- = \sqrt{\sum_{i=1}^{n} (\tilde{s}_{qi}(-)\tilde{\Lambda}_i^-)^\eta}$$
$$\cong (\sqrt{\sum_{i=1}^{n} \max(0,\ s_{qi1} - \Lambda_{i3}^-)^\eta},\ \sqrt{\sum_{i=1}^{n} (s_{qi2} - \Lambda_{i2}^-)^\eta},\ \sqrt{\sum_{i=1}^{n} (s_{qi3} - \Lambda_{i1}^-)^\eta}) \tag{4.46}$$

where η is a prespecified constant; $\eta \in Z^+$. When $\eta = 2$, it is equivalent to the traditional FTOPSIS method. When $\eta = 1$, it reduces to the fuzzy weighted average (FWA) method (Almutairi et al. 2022). Finally, the fuzzy closeness of location q is evaluated as

$$\tilde{C}_q = \frac{\xi \tilde{d}_q^-}{(1-\xi)\tilde{d}_q^+(+)\xi \tilde{d}_q^-} \tag{4.47}$$

Therefore,

$$C_{q1} = \min(\frac{\xi d_{q1}^-}{(1-\xi)d_{q3}^+ + \xi d_{q1}^-}, \frac{\xi d_{q3}^-}{(1-\xi)d_{q3}^+ + \xi d_{q3}^-}) \tag{4.48}$$

$$C_{q2} = \frac{\xi d_{q2}^-}{(1-\xi)d_{q2}^+ + \xi d_{q2}^-} \tag{4.49}$$

$$C_{q3} = \max(\frac{\xi d_{q1}^-}{(1-\xi)d_{q1}^+ + \xi d_{q1}^-}, \frac{\xi d_{q3}^-}{(1-\xi)d_{q1}^+ + \xi d_{q3}^-}) \tag{4.50}$$

where ξ is a prespecified constant; $\xi \in [0, 1]$. When $\xi = 0.5$, it reduces to the traditional FTOPSIS method. When $\xi = 0$, it is equivalent to a special case of fuzzy vise kriterijumska optimizacija i kompromisno (FVIKOR) (Wang et al. 2023b; Lin and Chen 2022). $0 \leq \tilde{C}_q \leq 1$. A location is more suitable for building the localized wafer fab if its fuzzy closeness is higher.

To get an absolute ranking, the fuzzy closeness can be defuzzified using the center-of-gravity (COG) method as (Motameni et al. 2019; Wang et al. 2021)

$$C_{q3} = \max(\frac{\xi d_{q1}^-}{(1-\xi)d_{q1}^+ + \xi d_{q1}^-}, \frac{\xi d_{q3}^-}{(1-\xi)d_{q1}^+ + \xi d_{q3}^-}) \tag{4.51}$$

Example 4.4 After the decision maker determined the country to sit the localized wafer fab, five factors are considered when selecting the appropriate location:

- Successful experiences;
- Cost of living;
- Personal and corporate taxes: Fr example, Samsung considered building new wafer fabs in three states—Texas, Arizona and New York. The top marginal state

income tax rates at these states are 0%, 5.1% and 11.7% (Wikipedia 2024b). For reference, Japan's income tax rates are much higher than the federal income tax rates in the United States, as shown in Table 4.11 (PWC, Individual—Taxes on personal income 2024; IRS, Federal income tax rates and brackets 2024).

- Available land: The localization of semiconductor supply chains has brought manufacturing back to the countries where the chips are designed, resulting in a significant impact on that country's supply and demand of industrial land. In the United States, for example, manufacturing-led land development accounted for 7–8% between 2018 and 2021. In 2022 and 2023, it jumped to more than 13%, and so far in 2024 it has accounted for 16.1% (e Costa and Vansnick 1997). Most are used for semiconductor manufacturing to allow more advanced manufacturers to open wafer fabs in the United States. Top 10 cities in the United States with available industrial land for building localized wafer fabs are summarized in Table 4.12 (e Costa and Vansnick 1997), among which the top three cities are located in Arizona, Texas and Missouri, further enhancing the motivation of building localized wafer fabs in these states. The personal and corporate taxes and avail lands in the three states are jointly compared in Fig. 4.12.
- Regulatory hurdles: Regulatory hurdles span a variety of types, such as legal, financial, compliance, and licensing. Failing to navigate these hurdles can lead to

Table 4.11 Comparison of Japan's and U.S.'s income tax rates (Data source: (PWC, Individual—Taxes on personal income 2024; IRS, Federal income tax rates and brackets 2024))

(Japan)

Income range (JPY)	in USD	Tax rate (%)
0 ~ 1,950,000	0 ~ 12,759	5%
1,950,001 ~ 3,300,000	12,759 ~ 21,592	10
3,300,001 ~ 6,950,000	21,592 ~ 45,473	20
6,950,001 ~ 9,000,000	45,473 ~ 58,886	23
9,000,001 ~ 18,000,000	58,886 ~ 117,772	33
18,000,001 ~ 40,000,000	117,772 ~ 261,717	40
40,000,001 ~	261,717 ~	45

(U.S.)

Income range (USD)	Tax rate (%)
0 ~ 11,000	10
11,001 ~ 44,725	12
44,726 ~ 95,375	22
95,376 ~ 182,100	24
182,101 ~ 231,250	32
231,251 ~ 578,125	35
578,126 ~	37

Table 4.12 Top 10 cities in the United States with available industrial land or building localized wafer fabs (Data source: CommercialEdge (e Costa and Vansnick 1997))

#	City	Area (Million Sq. Ft.)
1	Phoenix	39.07
2	Dallas-Ft. Worth	15.73
3	Kansas City	13.23
4	Atlanta	10.86
5	Chicago	10.26
6	Memphis	10.00
7	Columbus	9.89
8	Charlotte	9.09
9	Central Valley	8.92
10	Philadelphia	8.88

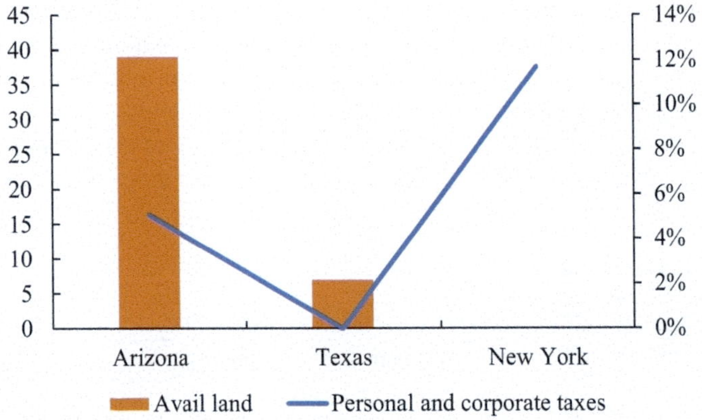

Fig. 4.12 Comparison of the personal and corporate taxes and avail lands in the three states

severe consequences, including legal penalties, reputational damage, and financial losses (JENKS2026, Navigating regulatory hurdles 2024). In the semiconductor industry, against the backdrop of wafer shortages and increased semiconductor merger and acquisition (M&A) activities, cross-border mergers and acquisitions and foreign direct investment (FDI) have come under increasing scrutiny globally, primarily due to competition, national security and industrial policy considerations. Regulators are particularly strict when it comes to advanced or sensitive technologies, and semiconductor mergers and acquisitions fall squarely within the scope of the regulatory regime (CommercialEdge 2024). TSMC's wafer fab localization plan reflects this consideration by choosing to partner with Japanese electronics giant Sony and auto parts maker Denso rather than owning its own wafer fabs in Japan (Kageyama 2024). In addition, taking a proactive approach to compliance is critical for businesses. This includes conducting regular risk assessments, implementing a strong compliance program and keeping abreast of regulatory changes. By effectively managing risks and compliance, businesses can more effectively navigate regulatory hurdles and reduce potential problems. Businesses can also benefit greatly from early engagement with regulators and building strong relationships. Collaboration can provide a greater understanding of regulatory expectations and help businesses stay informed of upcoming changes. Successful collaboration between businesses and regulators can improve compliance outcomes and streamline processes.

The following fuzzy judgment matrix is constructed by the decision maker by making pairwise comparisons:

$$\tilde{\mathbf{A}} = \begin{bmatrix} 1 & (1,\ 3,\ 5) & 1/(1,\ 3,\ 5) & (3,\ 5,\ 7) & (3,\ 5,\ 7) \\ 1/(1,\ 3,\ 5) & 1 & 1/(1,\ 3,\ 5) & 1/(1,\ 3,\ 5) & (2,\ 4,\ 6) \\ (1,\ 3,\ 5) & (1,\ 3,\ 5) & 1 & (2,\ 4,\ 6) & (3,\ 5,\ 7) \\ 1/(3,\ 5,\ 7) & (1,\ 3,\ 5) & 1/(2,\ 4,\ 6) & 1 & (3,\ 5,\ 7) \\ 1/(3,\ 5,\ 7) & 1/(2,\ 4,\ 6) & 1/(3,\ 5,\ 7) & 1/(3,\ 5,\ 7) & 1 \end{bmatrix}$$

A XAI tool, a group gradient bar chart, can be drawn to illustrated the pairwise comparison results, as shown in Fig. 4.13.

The fuzzy consistency ratio is about 0.14, which reveals the complexity of the decision-making problem. The fuzzy priorities of criteria are derived from the fuzzy judgment matrix using SCDT. For example,

$$\begin{aligned} \tilde{w}_1(\text{SCDT}) &= \tilde{w}_1(\text{FGM}) \cdot \frac{w_1(\text{CRISP})}{w_{12}(\text{FGM})} \\ &= (0.133,\ 0.287,\ 0.529) \cdot \frac{0.294}{0.287} \\ &= (0.136,\ 0.294,\ 0.542) \end{aligned}$$

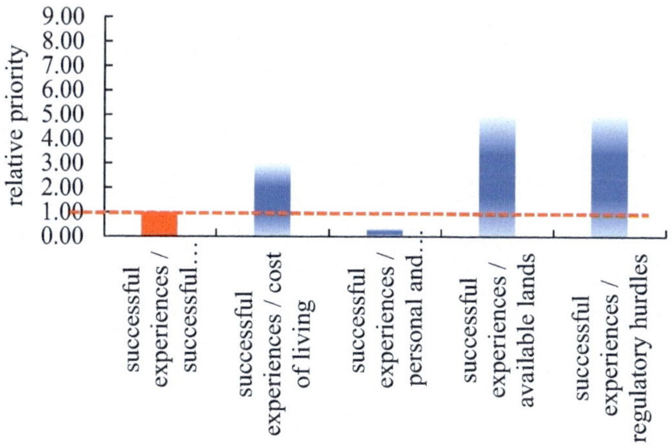

Fig. 4.13 Visualizing the pairwise comparison results

since $\sum_{i=1}^{n} |w_{12}(\text{FICSM}) - w_1(\text{CRISP})| \geq \sum_{i=1}^{n} |w_{12}(\text{FGM}) - w_1(\text{CRISP})|$. The derivation results are shown in Fig. 4.14. Obviously, the importance levels of the factors can be ranked as.

Personal and corporate taxes > Successful experiences > Available land > Cost of living > Regulatory hurdles.

Example 4.5 Based on the derived fuzzy priorities, eight possible locations in various regions with unequal conditions, are considered for building the localized wafer fab by the decision maker. The performances of these locations in optimizing various criteria are evaluated, and the evaluation results are presented in Table 4.13.

FTOPSIS is applied to evaluate and compare the overall performances of the eight locations for building the localized wafer fab. $\eta = 2$ and $\xi = 1$. First, the performance of a location in optimizing each criterion is normalized using fuzzy distributive normalization. For example,

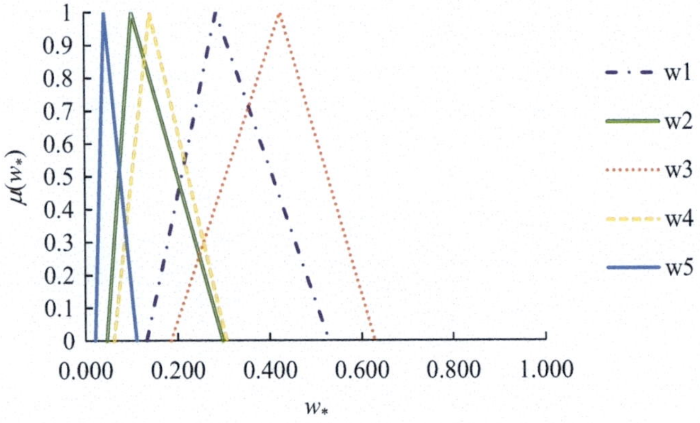

Fig. 4.14 Derivation results

Table 4.13 Performances of various locations

q	$\tilde{p}_{q1}$	$\tilde{p}_{q2}$	$\tilde{p}_{q3}$	$\tilde{p}_{q4}$	$\tilde{p}_{q5}$
1	(4, 5, 5)	(4, 5, 5)	(0, 1, 2)	(4, 5, 5)	(4, 5, 5)
2	(4, 5, 5)	(4, 5, 5)	(1, 2, 3)	(4, 5, 5)	(1.5, 2.5, 3.5)
3	(3, 4, 5)	(3, 4, 5)	(0, 1, 2)	(3, 4, 5)	(4, 5, 5)
4	(3, 4, 5)	(4, 5, 5)	(0, 1, 2)	(1, 2, 3)	(4, 5, 5)
5	(3, 4, 5)	(1.5, 2.5, 3.5)	(1, 2, 3)	(4, 5, 5)	(4, 5, 5)
6	(3, 4, 5)	(4, 5, 5)	(0, 1, 2)	(4, 5, 5)	(3, 4, 5)
7	(1.5, 2.5, 3.5)	(1, 2, 3)	(1.5, 2.5, 3.5)	(3, 4, 5)	(3, 4, 5)
8	(1.5, 2.5, 3.5)	(4, 5, 5)	(3, 4, 5)	(1, 2, 3)	(3, 4, 5)

$$\tilde{\rho}_{11} = \frac{1}{\sqrt{1 + \sum_{\phi \neq 1} (\frac{\tilde{p}_{\phi 1}}{\tilde{p}_{11}})^2}}$$
$$= \frac{1}{\sqrt{1 + (\frac{(4,\ 5,\ 5)}{(4,\ 5,\ 5)})^2 + (\frac{(3,\ 4,\ 5)}{(4,\ 5,\ 5)})^2 + \ldots + (\frac{(1.5,\ 2.5,\ 3.5)}{(4,\ 5,\ 5)})^2}}$$
$$= (0.311,\ 0.445,\ 0.554)$$

The normalization results are summarized in Table 4.14.

Subsequently, the fuzzy weighted scores of all locations are calculated based on the derived fuzzy priorities. For example,

$$\tilde{s}_{11} = \tilde{w}_1(\times)\tilde{\rho}_{11}$$
$$= (0.136,\ 0.294,\ 0.542)(\times)(0.311,\ 0.445,\ 0.554)$$
$$= (0.042,\ 0.131,\ 0.3)$$

The calculation results are summarized in Table 4.15.

Based on the fuzzy weighted scores, fuzzy ideal point and fuzzy anti-ideal point are specified respectively and presented in Table 4.16.

Subsequently, the distances from each location to the two reference points are measured, respectively. $\eta = 2$. The measurement results are summarized in Table 4.17. Finally, the fuzzy closeness of each location is evaluated. $\xi = 0.5$. The evaluation results are also shown in Table 4.17.

Table 4.14 Normalized performances of various locations

q	$\tilde{\rho}_{q1}$	$\tilde{\rho}_{q2}$	$\tilde{\rho}_{q3}$	$\tilde{\rho}_{q4}$	$\tilde{\rho}_{q5}$
1	(0.311, 0.445, 0.554)	(0.314, 0.407, 0.497)	(0, 0.171, 0.482)	(0.317, 0.423, 0.518)	(0.3, 0.403, 0.494)
2	(0.311, 0.445, 0.554)	(0.314, 0.407, 0.497)	(0.126, 0.342, 0.651)	(0.317, 0.423, 0.518)	(0.113, 0.201, 0.344)
3	(0.238, 0.356, 0.531)	(0.241, 0.325, 0.481)	(0, 0.171, 0.482)	(0.243, 0.338, 0.5)	(0.3, 0.403, 0.494)
4	(0.238, 0.356, 0.531)	(0.314, 0.407, 0.497)	(0, 0.171, 0.482)	(0.079, 0.169, 0.313)	(0.3, 0.403, 0.494)
5	(0.238, 0.356, 0.561)	(0.118, 0.203, 0.377)	(0.126, 0.342, 0.65)	(0.317, 0.423, 0.57)	(0.3, 0.403, 0.518)
6	(0.238, 0.356, 0.531)	(0.314, 0.407, 0.497)	(0, 0.171, 0.482)	(0.317, 0.423, 0.518)	(0.229, 0.322, 0.478)
7	(0.117, 0.222, 0.391)	(0.078, 0.163, 0.327)	(0.192, 0.427, 0.927)	(0.243, 0.338, 0.502)	(0.229, 0.322, 0.499)
8	(0.117, 0.222, 0.385)	(0.314, 0.407, 0.497)	(0.404, 0.683, 0.925)	(0.079, 0.169, 0.313)	(0.229, 0.322, 0.478)

Table 4.15 Fuzzy prioritized scores of various locations

q	$\tilde{s}_{q1}$	$\tilde{s}_{q2}$	$\tilde{s}_{q3}$	$\tilde{s}_{q4}$	$\tilde{s}_{q5}$
1	(0.042, 0.131, 0.3)	(0.014, 0.04, 0.141)	(0, 0.072, 0.301)	(0.02, 0.061, 0.161)	(0.006, 0.017, 0.054)
2	(0.042, 0.131, 0.3)	(0.014, 0.04, 0.141)	(0.023, 0.144, 0.406)	(0.02, 0.061, 0.161)	(0.002, 0.009, 0.037)
3	(0.033, 0.104, 0.288)	(0.011, 0.032, 0.137)	(0, 0.072, 0.301)	(0.015, 0.049, 0.156)	(0.006, 0.017, 0.054)
4	(0.033, 0.104, 0.288)	(0.014, 0.04, 0.141)	(0, 0.072, 0.301)	(0.005, 0.024, 0.097)	(0.006, 0.017, 0.054)
5	(0.033, 0.104, 0.304)	(0.005, 0.02, 0.107)	(0.023, 0.144, 0.406)	(0.02, 0.061, 0.177)	(0.006, 0.017, 0.056)
6	(0.033, 0.104, 0.288)	(0.014, 0.04, 0.141)	(0, 0.072, 0.301)	(0.02, 0.061, 0.161)	(0.005, 0.014, 0.052)
7	(0.016, 0.065, 0.212)	(0.003, 0.016, 0.093)	(0.035, 0.18, 0.579)	(0.015, 0.049, 0.156)	(0.005, 0.014, 0.054)
8	(0.016, 0.065, 0.209)	(0.014, 0.04, 0.141)	(0.074, 0.288, 0.577)	(0.005, 0.024, 0.097)	(0.005, 0.014, 0.052)

Table 4.16 Fuzzy ideal point and fuzzy anti-ideal point in Example 4.5

Reference Point	$\tilde{\Lambda}_1^{+/-}$	$\tilde{\Lambda}_2^{+/-}$	$\tilde{\Lambda}_3^{+/-}$	$\tilde{\Lambda}_4^{+/-}$	$\tilde{\Lambda}_5^{+/-}$
Fuzzy ideal point	(0.042, 0.131, 0.304)	(0.014, 0.04, 0.141)	(0.074, 0.288, 0.579)	(0.02, 0.061, 0.177)	(0.006, 0.017, 0.056)
Fuzzy anti-ideal point	(0, 0, 0)	(0, 0, 0)	(0, 0, 0)	(0, 0, 0)	(0, 0, 0)

Table 4.17 Measurement and evaluation results

q	$\tilde{d}_q^+$	$\tilde{d}_q^-$	$\tilde{C}_q$
1	(0, 0.216, 0.668)	(0.049, 0.167, 0.479)	(0.069, 0.435, 1)
2	(0, 0.144, 0.649)	(0.054, 0.208, 0.55)	(0.077, 0.59, 1)
3	(0, 0.218, 0.674)	(0.038, 0.141, 0.468)	(0.053, 0.392, 1)
4	(0, 0.221, 0.676)	(0.036, 0.136, 0.454)	(0.051, 0.382, 1)
5	(0, 0.148, 0.654)	(0.045, 0.19, 0.551)	(0.065, 0.562, 1)
6	(0, 0.218, 0.672)	(0.041, 0.147, 0.471)	(0.057, 0.403, 1)
7	(0, 0.129, 0.653)	(0.042, 0.199, 0.645)	(0.06, 0.606, 1)
8	(0, 0.075, 0.621)	(0.077, 0.3, 0.64)	(0.111, 0.8, 1)

Table 4.18 Defuzzification and ranking results

q	Defuzzified Closeness	Rank
1	0.501	5
2	0.556	2
3	0.482	7
4	0.478	8
5	0.542	4
6	0.487	6
7	0.556	3
8	0.637	1

Subsequently, the COG method is applied to defuzzify the fuzzy closeness of each location. The defuzzification results are summarized in Table 4.18.

According to the experimental results,

- Out of the eight locations, the eighth location has the highest overall performance for building the localized wafer fab, apparently due to its higher local government support while requiring less total investment. However, as mentioned earlier, a final decision may come after political negotiations. Location #2 comes in second.
- In contrast, location #4 is considered the least suitable due to the insufficient number of third parties in the global semiconductor chain at this location.

Example 4.6 In this example, FAHP-FTOPSIS is applied to prioritize possible measures to overcome the challenges of semiconductor supply chain localization. It is organized into a three-level decision hierarchy, as shown in Fig. 4.15. Nevertheless, each factor can be further decomposed into its sub-factors.

To apply FAHP, the relative weights of these factors are first compared pairwise in linguistic terms. The results are summarized in Table 4.19.

The following fuzzy judgment matrix is constructed to summarize the pairwise comparison results in Table 4.19:

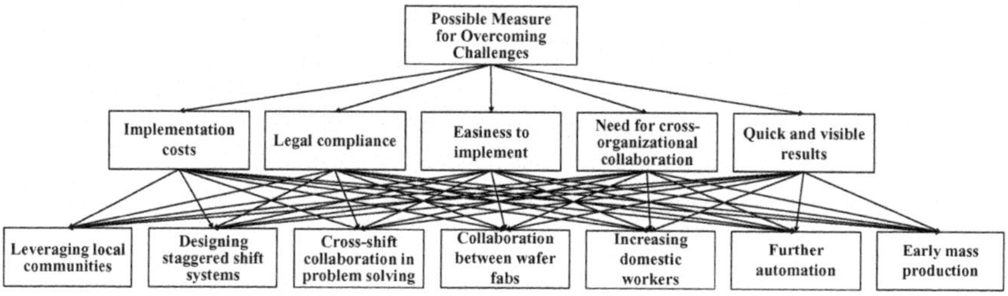

Fig. 4.15 Decision hierarchy

Table 4.19 Pairwise comparison results

Factor #1	Relative Weight (Factor #1 Over Factor #2)	Factor #2
Legal compliance	Strongly more important than	Implementation costs
Easiness to implement	Weakly more important than	Implementation costs
Need for cross-organization collaboration	Weakly or strongly more important than	Implementation costs
Implementation costs	Weakly more important than	Quick and visible results
Legal compliance	Weakly or strong more important than	Easiness to implement
Legal compliance	As equal as	Need for cross-organization collaboration
Legal compliance	Weakly more important than	Quick and visible results
Easiness to implement	As equal as or weakly more important than	Need for cross-organization collaboration
Quick and visible results	As equal as	Easiness to implement
Quick and visible results	Weakly more important than	Need for cross-organization collaboration

$$\tilde{\mathbf{A}} = \begin{bmatrix} 1 & 1/(3,\ 5,\ 7) & 1/(1,\ 3,\ 5) & 1/(2,\ 4,\ 6) & 1/(1,\ 3,\ 5) \\ (3,\ 5,\ 7) & 1 & (2,\ 4,\ 6) & (1,\ 1,\ 3) & (1,\ 3,\ 5) \\ (1,\ 3,\ 5) & 1/(3,\ 5,\ 7) & 1 & (1,\ 2,\ 4) & 1/(3,\ 5,\ 7) \\ (2,\ 4,\ 6) & 1/(1,\ 1,\ 3) & 1/(1,\ 2,\ 4) & 1 & (1,\ 3,\ 5) \\ (1,\ 3,\ 5) & 1/(1,\ 3,\ 5) & (1,\ 1,\ 3) & 1/(3,\ 5,\ 7) & 1 \end{bmatrix}$$

The fuzzy consistency ratio is about $0.97 < 0.1$, supporting the consistency among the pairwise comparison results. The fuzzy weights of factors are derived using the calibrated FGM (cFGM) (Chen et al. 2021), which calibrates the results of FGM to ensure the correctness in deriving the cores of fuzzy weights. The derivation results are shown in Fig. 4.16.

Based on the derived fuzzy weights, the seven possible measures for overcoming challenges in semiconductor supply chain localization are evaluated and compared using FTOPSIS. Beforehand, the performances of the possible measures in optimizing the five criteria have been estimated, as shown in Table 4.20.

First, the original performances are normalized using fuzzy distributive normalization. The normalization results are summarized in Table 4.21.

Subsequently, the fuzzy weighted scores of all possible measures are calculated and summarized in Table 4.22, for which the fuzzy weights of criteria are derived using FAHP.

Fuzzy ideal point and fuzzy anti-ideal point are then defined, respectively, from the fuzzy weighted scores (see Table 4.23).

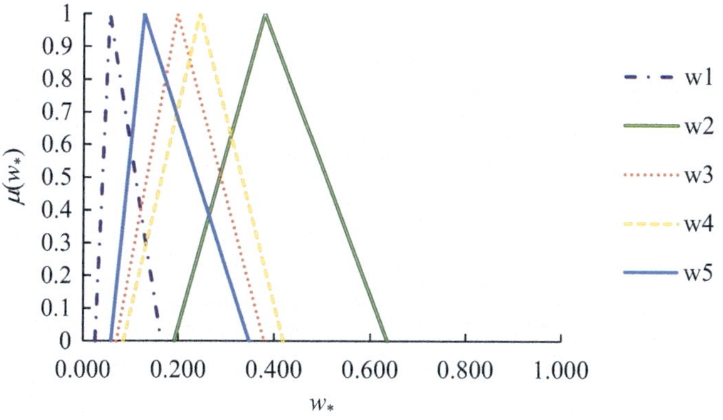

Fig. 4.16 Derived fuzzy weights of the factors

Table 4.20 Performances of the seven possible measures in optimizing the five criteria

q	Implementation costs	Legal compliance	Easiness to implement	Need for cross-organization collaboration	Quick and visible results
1	(3, 4, 5)	(3, 4, 5)	(1.5, 2.5, 3.5)	(1, 2, 3)	(3, 4, 5)
2	(1.5, 2.5, 3.5)	(0, 1, 2)	(3, 4, 5)	(1.5, 2.5, 3.5)	(3, 4, 5)
3	(3, 4, 5)	(3, 4, 5)	(1, 2, 3)	(1.5, 2.5, 3.5)	(1.5, 2.5, 3.5)
4	(1.5, 2.5, 3.5)	(4, 5, 5)	(1.5, 2.5, 3.5)	(1.5, 2.5, 3.5)	(1.5, 2.5, 3.5)
5	(1.5, 2.5, 3.5)	(1.5, 2.5, 3.5)	(3, 4, 5)	(4, 5, 5)	(4, 5, 5)
6	(0, 1, 2)	(4, 5, 5)	(0, 1, 2)	(4, 5, 5)	(1, 2, 3)
7	(0, 1, 2)	(4, 5, 5)	(1, 2, 3)	(4, 5, 5)	(3, 4, 5)

Table 4.21 Normalized performances

q	$\tilde{\rho}_{q1}$	$\tilde{\rho}_{q2}$	$\tilde{\rho}_{q3}$	$\tilde{\rho}_{q4}$	$\tilde{\rho}_{q5}$
1	(0.338, 0.551, 0.783)	(0.268, 0.374, 0.545)	(0.161, 0.342, 0.596)	(0.094, 0.202, 0.376)	(0.277, 0.423, 0.623)
2	(0.163, 0.344, 0.594)	(0, 0.094, 0.235)	(0.334, 0.547, 0.786)	(0.143, 0.253, 0.432)	(0.277, 0.423, 0.623)
3	(0.338, 0.551, 0.783)	(0.268, 0.374, 0.545)	(0.106, 0.273, 0.526)	(0.143, 0.253, 0.432)	(0.135, 0.264, 0.458)
4	(0.163, 0.344, 0.594)	(0.348, 0.468, 0.569)	(0.161, 0.342, 0.596)	(0.143, 0.253, 0.432)	(0.135, 0.264, 0.458)
5	(0.163, 0.344, 0.594)	(0.131, 0.234, 0.444)	(0.334, 0.547, 0.786)	(0.378, 0.506, 0.716)	(0.358, 0.529, 0.665)
6	(0, 0.138, 0.373)	(0.348, 0.468, 0.569)	(0, 0.137, 0.375)	(0.378, 0.506, 0.621)	(0.089, 0.211, 0.399)
7	(0, 0.138, 0.373)	(0.348, 0.468, 0.569)	(0.106, 0.273, 0.526)	(0.378, 0.506, 0.621)	(0.277, 0.423, 0.623)

Table 4.22 Fuzzy weighted scores

q	$\tilde{s}_{q1}$	$\tilde{s}_{q2}$	$\tilde{s}_{q3}$	$\tilde{s}_{q4}$	$\tilde{s}_{q5}$
1	(0.008, 0.03, 0.129)	(0.051, 0.142, 0.346)	(0.011, 0.067, 0.226)	(0.008, 0.049, 0.158)	(0.016, 0.054, 0.217)
2	(0.004, 0.019, 0.098)	(0, 0.035, 0.15)	(0.023, 0.107, 0.298)	(0.012, 0.061, 0.181)	(0.016, 0.054, 0.217)
3	(0.008, 0.03, 0.129)	(0.051, 0.142, 0.346)	(0.007, 0.054, 0.2)	(0.012, 0.061, 0.181)	(0.008, 0.034, 0.16)
4	(0.004, 0.019, 0.098)	(0.067, 0.177, 0.362)	(0.011, 0.067, 0.226)	(0.012, 0.061, 0.181)	(0.008, 0.034, 0.16)
5	(0.004, 0.019, 0.098)	(0.025, 0.089, 0.282)	(0.023, 0.107, 0.298)	(0.032, 0.123, 0.3)	(0.021, 0.067, 0.232)
6	(0, 0.008, 0.061)	(0.067, 0.177, 0.362)	(0, 0.027, 0.142)	(0.032, 0.123, 0.261)	(0.005, 0.027, 0.139)
7	(0, 0.008, 0.061)	(0.067, 0.177, 0.362)	(0.007, 0.054, 0.2)	(0.032, 0.123, 0.261)	(0.016, 0.054, 0.217)

Table 4.23 Fuzzy ideal point and fuzzy anti-ideal point in Example 4.6

Reference Point	$\tilde{\Lambda}_1^{+/-}$	$\tilde{\Lambda}_2^{+/-}$	$\tilde{\Lambda}_3^{+/-}$	$\tilde{\Lambda}_4^{+/-}$	$\tilde{\Lambda}_5^{+/-}$
Fuzzy ideal point	(0.008, 0.03, 0.129)	(0.067, 0.177, 0.362)	(0.023, 0.107, 0.298)	(0.032, 0.123, 0.3)	(0.021, 0.067, 0.232)
Fuzzy anti-ideal point	(0, 0.008, 0.061)	(0, 0.035, 0.15)	(0, 0.027, 0.142)	(0.012, 0.061, 0.181)	(0.005, 0.027, 0.139)

Subsequently, the overall performance of each possible measure is evaluated by measuring and comparing the fuzzy distances from it to the two reference points. The evaluation results are presented in Table 4.24. The overall performances are also defuzzified using COG to rank these possible measures unambiguously.

Table 4.24 Evaluation results using FTOPSIS

q	$\tilde{d}_q^+$	$\tilde{d}_q^-$	$\tilde{o}_q$	$COG(\tilde{o}_q)$	Rank
1	(0, 0.092, 0.57)	(0, 0.119, 0.504)	(0, 0.564, 1)	0.521	5
2	(0, 0.155, 0.593)	(0, 0.086, 0.441)	(0, 0.355, 1)	0.452	7
3	(0, 0.095, 0.573)	(0, 0.112, 0.478)	(0, 0.541, 1)	0.514	6
4	(0, 0.082, 0.564)	(0, 0.148, 0.494)	(0, 0.645, 1)	0.548	2
5	(0, 0.089, 0.566)	(0, 0.122, 0.559)	(0, 0.577, 1)	0.526	4
6	(0, 0.093, 0.562)	(0, 0.154, 0.484)	(0, 0.624, 1)	0.541	3
7	(0, 0.06, 0.554)	(0, 0.159, 0.53)	(0, 0.727, 1)	0.576	1

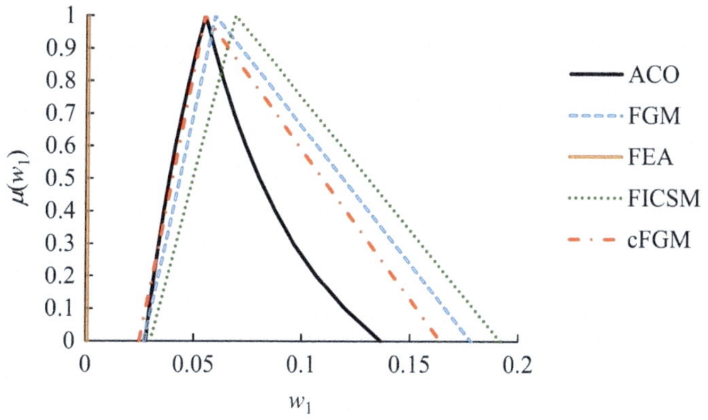

Fig. 4.17 Comparing the fuzzy weights derived using FGM, cFGM and ACO

According to the experimental results, the following discussion is made:

- To ensure that the fuzzy weights of the criteria estimated using cFGM were precise enough to generate a trustworthy decision, time-consuming ACO is also applied to derive the exact values of the fuzzy weights for comparison, against the results estimated by FGM, fuzzy extent analysis (FEA) and FICSM. The comparison results are shown in Fig. 4.17. Obviously, cFGM uses triangular fuzzy numbers to most precisely approximate the exact values of fuzzy weights to facilitate subsequent operations.
- The most important criterion for comparing possible measures to overcome semiconductor supply chain localization challenges is "legal compliance", which is not surprising since it is already in the execution/control phase and foundries should strive to minimize risks and run new wafer fabs well under way.
- The possible measure with the highest overall performance is "early mass production," which has the highest legal compliance and least need for cross-organizational collaboration, but suffers from higher implementation costs, consistent with Intel's recent move to sacrifice short-term competitiveness for future competitiveness and long-term sustainability.
- In contrast, "designing a staggered shift system" may be difficult to comply with legal requirements, so it ranks lowest.
- To validate the comparison results, four existing MCDM methods, fuzzy ordered weighted average (FOWA), FEA-FTOPSIS, FGM-FWA, and FGM-fuzzy Vise Kriterijumska Optimizacija I Kompromisno Resenje (FVIKOR) are also applied to this case. The ranking results using various methods are summarized in Fig. 4.18. As expected, the evaluation results using most of these methods are very close. "Early mass production" ranks first among all possible measures, except for "further automation" which is promoted to the first place by FOWA. TSMC's pursuit of full automation in its most advanced wafer fabs may one day also become a reality for localized wafer fabs in the United States.

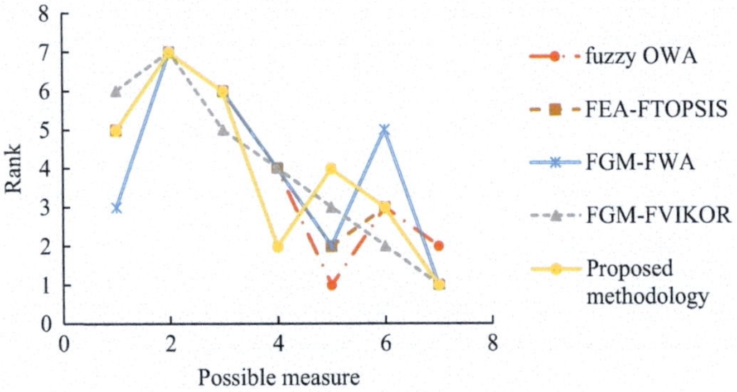

Fig. 4.18 Comparison of the ranking results using various methods

4.2.5 The FMACBETH-FVIKOR Approach

There are many MCDM methods that can be applied to choose the location for building the localized wafer fab. Therefore, the decisions made using different MCDM methods may not be the same, giving rise to an incentive to apply another MCDM method to confirm the choice made. To this end, the fuzzy measuring attractiveness by a categorical based evaluation technique (FMACBETH)-FVIKOR is introduced as follows.

The FMACBETH-FVIKOR approach is composed of the following steps (see Fig. 4.19):

Step 1. Structure the FMACBETH problem by building a value tree.

Step 2. Compare the relative attractivenesses of criteria in pairs: The results are put into a fuzzy judgement matrix.

Step 3. If the fuzzy judgement matrix is sufficiently consistent, calculate the fuzzy attractiveness of each criterion; otherwise, ask the decision maker to revise his/her judgements.

Step 4. (Optional) Perform a sensitivity analysis.

Step 5. Apply FVIKOR to evaluate and compare the overall performances of possible locations for building the localized wafer fab.

Step 6. Recommend the top performing locations.

First, MACBETH provides another way to derive the attractiveness (priority) of a criterion based on the pairwise comparison results (Chen 2021d). However, MACBETH uses an interval scale while AHP adopts a ratio scale. In addition, the calculation process of MACBETH is also different from that of AHP.

In FMACBETH, the decision maker constructs a consistent fuzzy judgment matrix $\tilde{\mathbf{A}} = \{\tilde{a}_{ij}\}$. $\tilde{a}_{ij}$, the relative attractiveness of criterion i with respect to criterion j, is mapped to Samejima et al. (2006); Chen and Wang 2021) instead (see Table 4.25) (Dhouib 2014). For example, if criterion i is very weakly more attractive than criterion j, $\tilde{a}_{ij} = (1,\ 2,\ 4)$.

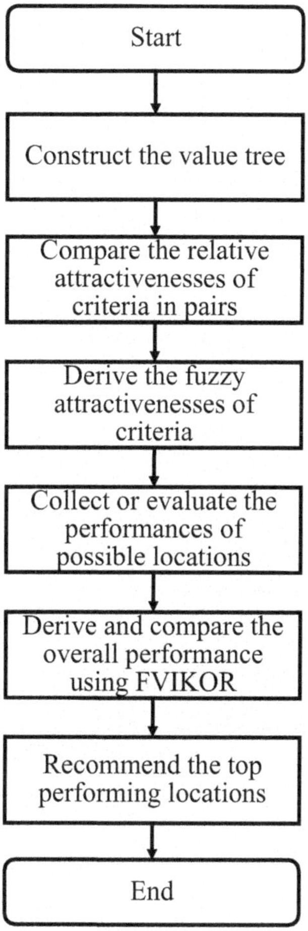

Fig. 4.19 Steps of the FMACBETH-FVIKOR approach

Table 4.25 Linguistic terms for representing the relative attractiveness

$\tilde{a}_{ij}$	Meaning
(1, 1, 3)	As attractive as
(1, 2, 4)	Very weakly more attractive than
(1, 3, 5)	Weakly more attractive than
(2, 4, 6)	Moderately more attractive than
(3, 5, 7)	Strongly more attractive than
(4, 6, 7)	Very strongly more attractive than
(5, 7, 7)	Extremely more attractive than

the most attractive → the least attractive

	Personal and corporate taxes	Successful experiences	Available land	Cost of living	Regulatory hurdles
(the most attractive) Personal and corporate taxes	No	Weak	Moderate	Weak	Strong
Successful experiences		No	Strong	Weak	Strong
Available land			No	Weak	Strong
Cost of living				No	Moderate
(the least attractive) Regulatory hurdles					No

Fig. 4.20 Fuzzy judgment matrix in FMACBETH

More attractive criteria will be placed closer to the upper left corner of $\tilde{\mathbf{A}}$ (see Fig. 4.20).

Therefore,

$$\tilde{w}_i \geq \tilde{w}_j \, \forall i < j \tag{4.52}$$

There are various ways to derive the fuzzy attractivenesses of criteria from $\tilde{\mathbf{A}}$ (Hanss 2005). For example, the following fuzzy quadratic programming (FQP) model can be optimized to fulfill this purpose:

(FQP Model)

$$\text{Min}\, \tilde{Z} = \sum_{i=1}^{n} \sum_{j \neq i} (\tilde{w}_i(-)\tilde{w}_j(-)(\tilde{a}_{ij} - 1)c)^2 \tag{4.53}$$

subject to

$$\tilde{w}_1 = 100 \tag{4.54}$$

$$\tilde{w}_n = 0 \tag{4.55}$$

$$0 \leq \tilde{w}_i \leq 100 \, \forall i \in [2, n-1] \tag{4.56}$$

$$c \geq 0 \tag{4.57}$$

The FQP problem needs to be converted into a crisp problem to be easily solved. First, according to the arithmetic for TFNs (Chen and Wang 2024; Opricovic 2011), the objective function is decomposed into

$$Z_1 = \sum_{i=1}^{n} \sum_{j \neq i} (w_{i1} - w_{j3} - a_{ij3}c + c)^2 \tag{4.58}$$

$$Z_2 = \sum_{i=1}^{n} \sum_{j \neq i} (w_{i2} - w_{j2} - a_{ij2}c + c)^2 \quad (4.59)$$

$$Z_3 = \sum_{i=1}^{n} \sum_{j \neq i} (w_{i3} - w_{j1} - a_{ij1}c + c)^2 \quad (4.60)$$

A common treatment is to defuzzify $\tilde{Z}$ using COG (Chen and Wang 2024) as

$$D(\tilde{Z}) = \frac{Z_1 + Z_2 + Z_3}{3} \quad (4.61)$$

In addition, Constraints (4.54) and (4.55) are fuzzily equivalent to

$$w_{11} = 50 \quad (4.62)$$

$$w_{12} = w_{13} = 100 \quad (4.63)$$

$$w_{n1} = w_{n2} = 0 \quad (4.64)$$

$$w_{n3} = 50 \quad (4.65)$$

Constraint (4.51) can be decomposed into

$$w_{i1} \geq 0 \quad (4.66)$$

$$w_{i3} \leq 100 \quad (4.67)$$

Finally, the following quadratic programming (QP) problem is solved instead:

(QP Model)

$$\text{Min}\, D(\tilde{Z}) = \frac{Z_1 + Z_2 + Z_3}{3} \quad (4.68)$$

subject to

$$Z_1 = \sum_{i=1}^{n} \sum_{j \neq i} (w_{i1} - w_{j3} - a_{ij3}c + c)^2 \quad (4.69)$$

$$Z_2 = \sum_{i=1}^{n} \sum_{j \neq i} (w_{i2} - w_{j2} - a_{ij2}c + c)^2 \quad (4.70)$$

$$Z_3 = \sum_{i=1}^{n} \sum_{j \neq i} (w_{i3} - w_{j1} - a_{ij1}c + c)^2 \tag{4.71}$$

$$w_{11} = 50 \tag{4.72}$$

$$w_{12} = w_{13} = 100 \tag{4.73}$$

$$w_{n1} = w_{n2} = 0 \tag{4.74}$$

$$w_{n3} = 50 \tag{4.75}$$

$$w_{i1} \geq 0 \forall i \in [2, n-1] \tag{4.76}$$

$$w_{i3} \leq 100 \forall i \in [2, n-1] \tag{4.77}$$

$$c \geq 0 \tag{4.78}$$

$$w_{i1} \leq w_{i2} \leq w_{i3} \forall i \tag{4.79}$$

Pairwise comparison results are inconsistent if $D(\tilde{Z}) \geq \xi$ where ξ is a threshold. The optimal solution needs to be normalized:

$$\begin{aligned} \tilde{w}_i &\to \frac{\tilde{w}_i}{\sum_{j=1}^{n} \tilde{w}_j} \\ &= \frac{1}{1 + \sum_{j \neq i} \frac{\tilde{w}_j}{\tilde{w}_i}} \end{aligned} \tag{4.80}$$

except that $w_{n1} = w_{n2} = 0$. Fuzzy attractivenesses derived by solving the QP Model may be symmetric. Nevertheless, after normalization, fuzzy attractivenesses become asymmetric.

Example 4.7 For the previous example, the fuzzy judgment matrix using FMACBETH is

$$\tilde{\mathbf{A}} = \begin{bmatrix} 1 & (1,\ 3,\ 5) & (2,\ 4,\ 6) & (1,\ 3,\ 5) & (3,\ 5,\ 7) \\ - & 1 & (3,\ 5,\ 7) & (1,\ 3,\ 5) & (3,\ 5,\ 7) \\ - & - & 1 & (1,\ 3,\ 5) & (3,\ 5,\ 7) \\ - & - & - & 1 & (2,\ 4,\ 6) \\ - & - & - & - & 1 \end{bmatrix}$$

The NP model for deriving the fuzzy attractivenesses of criteria is shown in Fig. 4.21. The optimal solution is

$$\begin{aligned} \tilde{w}_1^* &= (50,\ 100,\ 100) \\ \tilde{w}_2^* &= (61.1,\ 61.1,\ 61.1) \\ \tilde{w}_3^* &= (51.1,\ 51.1,\ 51.1) \\ \tilde{w}_4^* &= (48.9,\ 48.9,\ 48.9) \\ \tilde{w}_5^* &= (0,\ 0,\ 50) \end{aligned}$$

After normalization,

$$\begin{aligned} \tilde{w}_1^* &= (0.191, 0.383, 0.383) \\ \tilde{w}_2^* &= (0.196, 0.234, 0.289) \\ \tilde{w}_3^* &= (0.164, 0.196, 0.242) \\ \tilde{w}_4^* &= (0.157, 0.188, 0.232) \end{aligned}$$

```
min=DZ;
DZ=1/3*Z1+1/3*Z2+1/3*Z3;
Z1=(w11-w23-5*c+c)^2+(w11-w33-6*c+c)^2+(w11-w43-5*c+c)^2+(w11-w53-
7*c+c)^2+(w21-w33-7*c+c)^2+(w21-w43-5*c+c)^2+(w21-w53-7*c+c)^2+(w31-w43-
5*c+c)^2+(w31-w53-7*c+c)^2+(w41-w53-6*c+c)^2;
Z2=(w12-w22-3*c+c)^2+(w12-w32-4*c+c)^2+(w12-w42-3*c+c)^2+(w12-w52-
5*c+c)^2+(w22-w32-5*c+c)^2+(w22-w42-3*c+c)^2+(w22-w52-5*c+c)^2+(w32-w42-
3*c+c)^2+(w32-w52-5*c+c)^2+(w42-w52-4*c+c)^2;
Z3=(w13-w21-1*c+c)^2+(w13-w31-2*c+c)^2+(w13-w41-1*c+c)^2+(w13-w51-
3*c+c)^2+(w22-w31-3*c+c)^2+(w23-w41-1*c+c)^2+(w23-w51-3*c+c)^2+(w33-w41-
1*c+c)^2+(w33-w51-3*c+c)^2+(w43-w51-2*c+c)^2;
w11=50;
w12=100;
w13=100;
w51=0;
w52=0;
w53=50;
w11<=w12;
w12<=w13;
w21<=w22;
w22<=w23;
w31<=w32;
w32<=w33;
w41<=w42;
w42<=w43;
w51<=w52;
w52<=w53;
```

Fig. 4.21 NP model for deriving the fuzzy attractivenesses of criteria

$$\tilde{w}_5^* = (0, 0, 0.191)$$

Subsequently, FVIKOR (Fan et al. 2023; Chen et al. 2020) is applied to evaluate and compare the overall performance of each location for building the localized wafer fab. FVIKOR comprises the following steps:

Step 1. Determine the best and worst performances in optimizing each criterion:

$$\begin{aligned} p_i^* &= \max_h p_{hi} \\ &= (\max_h p_{hi1}, \max_h p_{hi2}, \max_h p_{hi3}) \end{aligned} \tag{4.81}$$

$$\begin{aligned} p_i^- &= \min_h p_{hi} \\ &= (\min_h p_{hi1}, \min_h p_{hi2}, \min_h p_{hi3}) \end{aligned} \tag{4.82}$$

where p_{hi} is the performance of location h for building the localized wafer fab in optimizing criterion i.; $h = 1 \sim H$. H is the number of locations considered by the decision maker.

Step 2. Compute the normalized fuzzy distance from each location for building the localized wafer fab to the best performance:

$$\begin{aligned} d_{hi} &= \frac{p_i^* - p_{hi}}{p_{i3}^* - p_{i1}^-} \\ &= (\frac{p_{i1}^* - p_{hi3}}{p_{i3}^* - p_{i1}^-}, \frac{p_{i2}^* - p_{hi2}}{p_{i3}^* - p_{i1}^-}, \frac{p_{i3}^* - p_{hi1}}{p_{i3}^* - p_{i1}^-}) \end{aligned} \tag{4.83}$$

Step 3. Compute the sum of the weighted distances S_h and the maximum weighted distance R_h (Lin and Chen 2022):

$$S_h = \sum_{i=1}^{n} (w_i d_{hi}) \tag{4.84}$$

$$R_h = \max_i (w_i d_{hi}) \tag{4.85}$$

S_h considers the performances in optimizing all criteria, while R_h highlights the performance in optimizing the most important criterion or the worse performance.

Step 4. Compute the value of Q_h by compromising between S_h and R_h (Lin and Chen 2022):

$$Q_h = \xi \cdot \frac{S_h - \min_r S_r}{\max(\max_r S_r) - \min(\min_r S_r)} (+) (1 - \xi) \cdot \frac{R_h - \min_r R_r}{\max(\max_r R_r) - \min(\min_r R_r)} \tag{4.86}$$

where $\xi \in [0, 1]$.

Step 5. Defuzzify S_h, R_h, Q_h in the following way (Lin and Chen 2022):

$$D(S_h) = \frac{S_{h1} + 2S_{h2} + S_{h3}}{4} \tag{4.87}$$

$$D(R_h) = \frac{R_{h1} + 2R_{h2} + R_{h3}}{4} \tag{4.88}$$

$$D(Q_h) = \frac{Q_{h1} + 2Q_{h2} + Q_{h3}}{4} \tag{4.89}$$

Step 6. Rank locations for building the localized wafer fab according to their $D(S_h)$, $D(R_h)$, and $D(Q_h)$ values from the smallest to the largest. The decision maker will have three ranking results, giving him/her high flexibility, which is an advantage of FVIKOR over FTOPSIS (Chen et al. 2023a, 2021; Nazar et al. 2019; Lin and Chen 2022). For example, when $D(Q_h)$ is considered, the top two locations for building the localized wafer fab are indicated with locations $h_{(1)}$ and$h_{(2)}$, respectively. Then, in the view of Opricovic (2011), alternative $h_{(1)}$ can be recommended to the decision maker if the following two conditions are met:

$$D(Q_{h_{(2)}}) - D(Q_{h_{(1)}}) \geq \frac{1}{H-1} \tag{4.90}$$

$$D(S_{h_{(1)}}) = \min_r D(S_r) \quad \text{or} \quad D(R_{h_{(1)}}) = \min_r D(R_r) \tag{4.91}$$

Example 4.8 For Example 4.5, the best and worst performances in optimizing each criterion are determined. The results are shown in Table 4.26.

The normalized fuzzy distance between each location for building the localized wafer fab and the best performance was measured. The measurement results are summarized in Table 4.27.

The values of S_h and R_h are then computed for each location for building the localized wafer fab. The results are summarized in Table 4.28. Based on them, the Q_h of the COVID-19 vaccination location was derived by setting ξ to 0.5.

The defuzzified values of these performance measures are summarized in Table 4.29. Based on the defuzzification results, various locations for building the localized wafer fab are ranked, as shown in Table 4.29. Location #2 achieves the

Table 4.26 Best and worst performances in optimizing each criterion

i	p_i^*	p_i^-
1	(4, 5, 5)	(1.5, 2.5, 3.5)
2	(3, 4, 5)	(0, 1, 2)
3	(4, 5, 5)	(1.5, 2.5, 3.5)
4	(4, 5, 5)	(1, 2, 3)
5	(4, 5, 5)	(1.5, 2.5, 3.5)

Table 4.27 Normalized fuzzy distance between each location for building the localized wafer fab and the best performance

h	d_{h1}	d_{h2}	d_{h3}	d_{h4}	d_{h5}
1	(0, 0, 0.29)	(0, 0, 0.25)	(0.2, 0.6, 1)	(0, 0, 0.25)	(0, 0, 0.29)
2	(0, 0, 0.29)	(0, 0, 0.25)	(0, 0.4, 0.8)	(0, 0, 0.25)	(0.14, 0.71, 1)
3	(0, 0.29, 0.57)	(0, 0.25, 0.5)	(0.2, 0.6, 1)	(0, 0.25, 0.5)	(0, 0, 0.29)
4	(0, 0.29, 0.57)	(0, 0, 0.25)	(0.2, 0.6, 1)	(0.25, 0.75, 1)	(0, 0, 0.29)
5	(0, 0.29, 0.57)	(0.13, 0.63, 0.88)	(0, 0.4, 0.8)	(0, 0, 0.25)	(0, 0, 0.29)
6	(0, 0.29, 0.57)	(0, 0, 0.25)	(0.2, 0.6, 1)	(0, 0, 0.25)	(0, 0.29, 0.57)
7	(0.14, 0.71, 1)	(0.25, 0.75, 1)	(0, 0.3, 0.7)	(0, 0.25, 0.5)	(0, 0.29, 0.57)
8	(0.14, 0.71, 1)	(0, 0, 0.25)	(0, 0, 0.4)	(0.25, 0.75, 1)	(0, 0.29, 0.57)

Table 4.28 S_h, R_h and Q_h of each location for building the localized wafer fab

h	S_h	R_h	Q_h
1	(0.04, 0.23, 0.64)	(0.04, 0.23, 0.38)	(0, 0.14, 0.81)
2	(0, 0.15, 0.7)	(0, 0.15, 0.31)	(0, 0, 0.74)
3	(0.04, 0.39, 0.83)	(0.04, 0.23, 0.38)	(0, 0.22, 0.91)
4	(0.08, 0.44, 0.89)	(0.04, 0.23, 0.38)	(0, 0.24, 0.94)
5	(0.02, 0.34, 0.79)	(0.02, 0.15, 0.31)	(0, 0.09, 0.79)
6	(0.04, 0.3, 0.77)	(0.04, 0.23, 0.38)	(0, 0.17, 0.88)
7	(0.07, 0.47, 1.01)	(0.04, 0.17, 0.29)	(0, 0.17, 0.88)
8	(0.07, 0.31, 0.84)	(0.04, 0.17, 0.29)	(0, 0.09, 0.79)

lowest value of $D(Q_h)$, followed by location #5. Namely, $h_{(1)} = 2$ and $h_{(2)} = 5$. However, $D(Q_{h_{(2)}}) - D(Q_{h_{(1)}}) = 0.242 - 0.186 = 0.056 \ngeq \frac{1}{8-1} = 0.143$. Therefore, the superiority of location #2 over location #5 is not significant. Such a conclusion is also slightly different from that using the FAHP-FTOPSIS approach in which location #8 ranks first, as compared in Table 4.30.

Table 4.29 Defuzzification results

h	$D(S_h)$	$D(R_h)$	$D(Q_h)$	Rank
1	0.283	0.220	0.272	4
2	0.251	0.153	0.186	1
3	0.414	0.220	0.336	7
4	0.461	0.220	0.355	8
5	0.370	0.158	0.242	2
6	0.351	0.220	0.306	5
7	0.505	0.166	0.307	6
8	0.381	0.166	0.245	3

Table 4.30 Comparison of the results using two MCDM methods

h	FAHP-FTOPSIS	FMACBETH-FVIKOR
1	5	4
2	2	1
3	7	7
4	8	8
5	4	2
6	6	5
7	3	6
8	1	3

4.2.6 Visualizing the Evaluation Results in FMACBETH-FVIKOR

A bidirectional scatterplot (Wang and Chen 2023) can also be applied to visualize the evaluation results in FMACBETH-FVIKOR (see Fig. 4.22):

- The perfect location is in the center, surrounded by all possible locations.

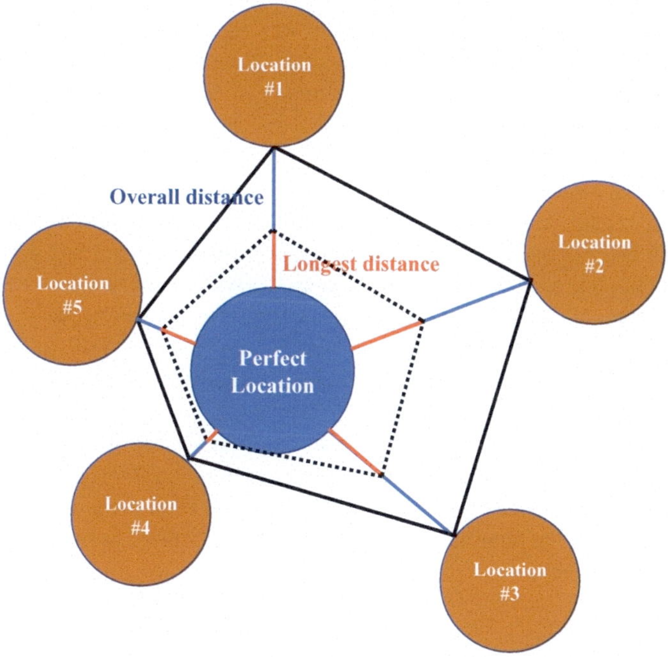

Fig. 4.22 Bidirectional scatterplot for visualizing the comparison results using FMACBETH-FVIKOR

- The longest distance and overall distance from each possible location to the perfect location are shown.
- The distances of various possible locations are connected through polygons to facilitate their comparison.

4.3 Qualitative Analysis in Conjunction with Quantitative Methods for Semiconductor Supply Chain Localization

4.3.1 Application of SWOT Analysis in Conjunction with Quantitative Methods

Almutairi et al. (2022) used SWOT analysis to solve Iran's energy problems. Because most of the country's energy sources are oil that causes significant harm to the environment, alternative energy sources must be found to protect the environment. Therefore, they evaluated Iran's renewable energy sources for energy development. A MCDM method was applied to rank the elements of the SWOT analysis. Finally, they drew the conclusion that Iran was most suitable for using wind and solar power generation.

Fan et al. (2023) studied the tourism marketing strategy of Changbai Mountain in China. In the study, SWOT analysis was used to list the strengths, weaknesses, opportunities and threats of Changbai Mountain and then create a suitable tourism marketing strategy. Subsequently, two MCDM methods, quantitative strategic planning matrix (QSPM) and AHP, were combined to prioritize strategies.

Zhu et al. (2023) also analyzed China's tourism industry in their study, which used SWOT analysis combined with TOPSIS to sort the strategic priorities of West Lake eco-tourism to analyze the current situation and estimate the future development.

Voukkali and Zorpas (2022) studied the relationship between urban development and environmental recycling in Europe, and used SWOT analysis and AHP to select urban development strategies to meet future urban development needs.

Kaymaz et al. (2022) also combined SWOT analysis and AHP to analyze the social and economic development of Erzurum, a Turkish city, to identify weaknesses and select new development goals.

Savari et al. (2022) combined SWOT analysis, FAHP (Chen 2023f, 2023g, 2023h, 2023i; Kubler et al. 2016; Ishizaka 2014; Naghadehi et al. 2009) and Dawes matrix (also known as situation analysis method, TOWS) to choose the most advantageous strategy among twelve alternatives to facilitate small farmers to plant and produce when the land was dry.

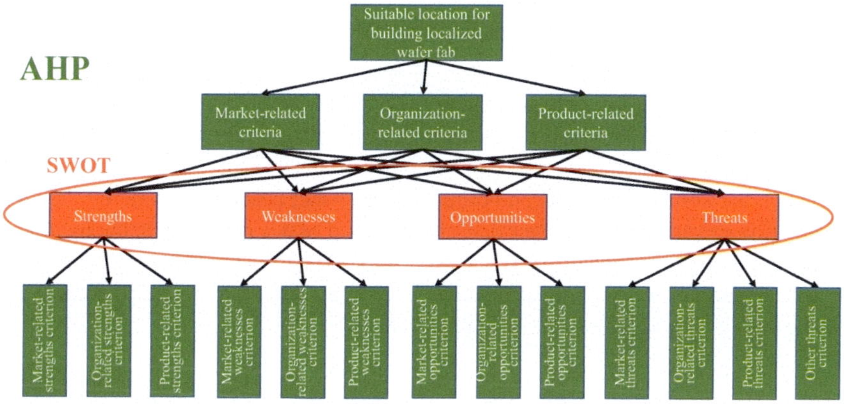

Fig. 4.23 Embedding SWOT into AHP

4.3.2 Embedding SWOT into AHP

Shinno et al. (2006) applied AHP to evaluate the global competitiveness of enterprises, in which influencing factors were divided into four categories, i.e., strengths, weaknesses, opportunities and threats, following the concept of SWOT analysis. This is the most direct approach for combining the advantages of AHP and SWOT, and can also be applied to make decisions in semiconductor supply chain localization, as shown in Fig. 4.23.

The combination of AHP and SWOT is reasonable because in AHP problems some criteria are the-higher-the-better (HTB), while others may be the-lower-the-better (LTB). Likewise, in SWOT analysis, strengths and opportunities are HTB factors, while weaknesses and threats are LTB factors.

4.3.3 Application of the SWOT-AHP-TOPSIS Approach for Semiconductor Supply Chain Localization

Before applying the AHP-TOPSIS approach, the performances of all locations for building the localized wafer fab have been collected. When two locations are compared, the better performance at one location becomes its advantage (or strength), while the poorer performance at the location becomes a disadvantage (or weakness), on any criterion. However, such a difference must be significant enough. Furthermore, for the consistency of baselines, it is better to compare the normalized values instead of the original values.

For example, in the previous case, the locations in the United States and Japan for building the localized wafer fab are to be compared. Assume that a deviation of at least 0.15 is sufficient to conclude that one has an advantage over the other. Then, according to Table 4.5,

- When considering the first, third, and fifth criteria, namely energy costs, natural disaster risks and technical talent adequacy, the United States has more advantages than Japan.
- In comparison, when the second criterion (i.e., water costs) is considered, Japan's location has an advantage over that of the United States for building the localized wafer fab.
- In the fourth criterion (human resource costs), the difference in the performances between the two locations is not significant.

The comparison results can be summarized as Fig. 4.24.

However, since these criteria have different priorities, it would be even better to compare the prioritized scores rather than the normalized performances. The problem is that the prioritized scores are usually very small values, so it is difficult to establish a unified threshold for the difference between two prioritized scores.

Strengths	**Weaknesses**
• Lower energy costs • Lower natural disaster risks • More adequate technical talents	• Higher human resource costs
Opportunities	**Threats**
-	-

(Localization Plan of TSMC in the U.S.)

Strengths	**Weaknesses**
• Lower human resource costs	• Higher energy costs • Higher natural disaster risks • Insufficient technical talents
Opportunities	**Threats**
-	-

(Localization Plan of TSMC in Japan)

Fig. 4.24 Comparison results using the SWOT-AHP-TOPSIS approach

For the previous case, the threshold for the difference between two prioritized scores is established as 0.03. According to Table 4.6,

- The location in the United States has an advantage over that in Japan for building the localized wafer fab when the first and fifth criteria (i.e., energy costs and technical talent adequacy) are considered.
- Japan's location also has an advantage over that of the United States for building the localized wafer fab when the second criterion (i.e., water costs) is considered.

The results are embedded into the SWOT matrixes in Fig. 4.25.

To compare the locations in all countries for building the localized wafer fab simultaneously, their prioritized scores are sorted, where the top three localized wafer fab locations with $s_{qi} \geq 0.05$ have an advantage over the remaining locations. Conversely, the bottom three locations with $s_{qi} \geq 0.05$ are at a disadvantage relative to the rest. The results are summarized in Fig. 4.26.

Strengths	**Weaknesses**
• Lower energy costs • More adequate technical talents	• Higher human resource costs
Opportunities	**Threats**
-	-

(Localization Plan of TSMC in the U.S.)

Strengths	**Weaknesses**
• Lower human resource costs	• Higher energy costs • Insufficient technical talents
Opportunities	**Threats**
-	-

(Localization Plan of TSMC in Japan)

Fig. 4.25 Comparison results based on the prioritized scores

Strengths	**Weaknesses**
• Lower energy costs • Lower human resource costs • More adequate technical talents	• Higher water costs
Opportunities	**Threats**
-	-

(Localization Plan of TSMC in the U.S.)

Strengths	**Weaknesses**
• Lower water costs • Lower human resource costs	• Higher energy costs
Opportunities	**Threats**
-	-

(Localization Plan of TSMC in Japan)

Strengths	**Weaknesses**
• Lower water costs • Higher political stability	• Higher energy costs
Opportunities	**Threats**
-	-

(Localization Plan of TSMC in Germany)

Strengths	**Weaknesses**
-	• Higher human resource costs
Opportunities	**Threats**
-	-

(Localization Plan of TSMC in France)

Fig. 4.26 Comparison of the locations in all countries for building the localized wafer fab simultaneously

Strengths	Weaknesses
-	• Higher energy costs • Higher water costs • Higher human resource costs
Opportunities	**Threats**
-	-

(Localization Plan of TSMC in the U.K.)

Strengths	Weaknesses
• Lower energy costs • Higher political stability	• Higher water costs • Higher human resource costs
Opportunities	**Threats**
-	-

(Localization Plan of TSMC in Canada)

Strengths	Weaknesses
• Lower energy costs • Lower water costs • Lower human resource costs	-
Opportunities	**Threats**
-	-

(Localization Plan of TSMC in South Korea)

Fig. 4.26 (continued)

For a wafer foundry, the opportunity to choose a location to build a localized wafer fab lies in the absence or slow progress of its current (or new) competitors building wafer fabs there. Conversely, there is a threat if its current (or new) competitors make faster progress in building wafer fabs locally. These conclusions are based on quantifiable data.

References

F. Ahmed, K. Kilic, Fuzzy analytic hierarchy process: a performance analysis of various algorithms. Fuzzy Sets Syst. **362**, 110–128 (2019)

K. Almutairi, S.J. Hosseini Dehshiri, S.S. Hosseini Dehshiri, A. Mostafaeipour, A.X. Hoa, K. Techato, Determination of optimal renewable energy growth strategies using SWOT analysis, hybrid MCDM methods, and game theory: a case study. Int. J. Energy Res. **46**(5), 6766–6789 (2022)

V. Batth, Toyota motor corporation: just in time (jit) management strategy or beyond. J. Case Res. **12**(1), 18–27 (2021)

E. Buber, D.I.R.I. Banu, Performance analysis and CPU vs GPU comparison for deep learning. in *2018 6th International Conference on Control Engineering and Information Technology* (2018), pp. 1–6

J.J. Buckley, Fuzzy hierarchical analysis. Fuzzy Sets Syst. **17**(3), 233–247 (1985)

Business Weekly, What's wrong with saying "goodbye" to colleagues after get off work? Japanese working in Taiwan: I was shocked (2024). https://www.businessweekly.com.tw/international/blog/3014886

T. Chen, Strengthening the competitiveness and sustainability of a semiconductor manufacturer with cloud manufacturing. Sustainability **6**(1), 251–266 (2014)

T. Chen, A diversified AHP-tree approach for multiple-criteria supplier selection. Comput. Manag. Sci. **18**, 431–453 (2021a)

T.C.T. Chen, Type-II fuzzy collaborative intelligence for assessing cloud manufacturing technology applications. Robot. Comput.-Integrat. Manuf. **78**, 102399 (2022)

T.C.T. Chen, M.C. Chiu, Evaluating the sustainability of smart technology applications in healthcare after the COVID-19 pandemic: a hybridising subjective and objective fuzzy group decision-making approach with explainable artificial intelligence. Digital Health **8**, 20552076221136380 (2022)

T.C.T. Chen, C.W. Lin, Assessing cloud manufacturing applications using an optimally rectified FAHP approach. Complex and Intell. Syst. **8**(6), 5087–5099 (2022a)

T.C.T. Chen, Y.C. Lin, Fuzzified deep neural network ensemble approach for estimating cycle time range. Appl. Soft Comput. **130**, 109697 (2022b)

T. Chen, Y.C. Wang, Interval fuzzy number-based approach for modeling an uncertain fuzzy yield learning process. J. Ambient. Intell. Humaniz. Comput. **11**, 1213–1223 (2020)

T.C.T. Chen, Y.C. Wang, A fuzzy mid-term capacity and production planning model for a manufacturing system with cloud-based capacity. Complex and Intell. Syst. **7**(1), 71–85 (2021)

T.C.T. Chen, Y.C. Wang, Evaluating innovative future robotic applications in manufacturing using a fuzzy collaborative intelligence approach. The Int. J. Adv. Manufact. Technol. **130**(11), 6027–6041 (2024)

T. Chen, Y.C. Wang, H.C. Wu, Analyzing the impact of vaccine availability on alternative supplier selection amid the COVID-19 pandemic: a cFGM-FTOPSIS-FWI approach. Healthcare **9**(1), 71 (2021)

T. Chen, Y.C. Wang, P.H. Jiang, A selectively calibrated derivation technique and generalized fuzzy TOPSIS for semiconductor supply chain localization assessment. Decision Anal. J. **8**, 100275 (2023a)

T.C.T. Chen, Y.C. Wang, M.C. Chiu, An efficient approximating alpha-cut operations approach for deriving fuzzy priorities in fuzzy multi-criterion decision-making. Appl. Soft Comput. **139**, 110238 (2023b)

T.C.T. Chen, H.C. Wu, M.C. Chiu, A deep neural network with modified random forest incremental interpretation approach for diagnosing diabetes in smart healthcare. Appl. Soft Comput. **152**, 111183 (2024a)

T.C.T. Chen, Deriving the priorities of criteria. in *Advances in Fuzzy Group Decision Making* (2021), pp. 29–53

T.C.T. Chen, Consensus measurement and enhancement. in *Advances in Fuzzy Group Decision Making* (2021). pp. 55–72.

T.C.T. Chen, Introduction to fuzzy group decision-making. in *Advances in Fuzzy group decision making* (2021), pp. 1–9

T.C.T. Chen, Defect pattern analysis, yield learning modeling, and yield prediction. in *Production Planning and Control in Semiconductor Manufacturing: Big Data Analytics and Industry 4.0 Applications* (2023), pp. 63–76

T.-C. T. Chen, Explainable artificial intelligence (XAI) in manufacturing. in *Explainable Artificial Intelligence (XAI) in Manufacturing: Methodology, Tools, and Applications* (2023), pp. 1–11

T.C.T. Chen, Applications of XAI for forecasting in the manufacturing domain. in *Explainable Artificial Intelligence (XAI) in Manufacturing: Methodology, Tools, and Applications* (2023), pp. 13–50

T.C.T. Chen, Big data analytics for semiconductor manufacturing. in *Production Planning and Control in Semiconductor Manufacturing: Big Data Analytics and Industry 4.0 Applications* (2023), pp. 1–19

T.C.T. Chen, Industry 4.0 for semiconductor manufacturing. in *Production Planning and Control in Semiconductor Manufacturing: Big Data Analytics and Industry 4.0 Applications* (2023), pp. 21–40

T.C.T. Chen, Applications of XAI for decision making in the manufacturing domain. in *Explainable Artificial Intelligence (XAI) in Manufacturing: Methodology, Tools, and Applications* (2023), pp. 51–81

T.C.T. Chen, Smart technology applications in healthcare before, during, and after the COVID-19 pandemic. in *Sustainable Smart Healthcare: Lessons Learned from the COVID-19 Pandemic* (2023), pp. 19–37

T.C.T. Chen, Sustainable smart healthcare applications: lessons learned from the COVID-19 pandemic. in *Sustainable Smart Healthcare: Lessons Learned from the COVID-19 Pandemic* (2023), pp. 65–92

T. C. T. Chen, Applications of XAI to job sequencing and scheduling in manufacturing. in *Explainable Artificial Intelligence (XAI) in Manufacturing: Methodology, Tools, and Applications* (2023), pp. 83–105

T. C. T. Chen, XAmI applications to telemedicine and telecare. in *Explainable Ambient Intelligence (XAmI) Explainable Artificial Intelligence Applications in Smart Life* (2024), pp. 85–109

T.C.T. Chen, Y.C. Wang, AI applications to pull production, JIT, and production leveling. in *Artificial Intelligence and Lean Manufacturing* (2022), pp. 55–74

T.C.T. Chen, Y.C. Wang, KI-anwendungen in pull-produktion, JIT und produktionsnivellierung. in *Künstliche Intelligenz und schlanke Produktion* (2023), pp. 59–79

T.C.T. Chen, W.A. Chaovalitwongse, M.J. O'grady, K. Honda, Editorial: smart technologies for improving the quality of mobile health care. Health Care Managem. Sci. **23**, 171–172 (2020)

T.C.T. Chen, Y.C. Wang, C.W. Lin, Enhancing the effectiveness of output projection in wafer fabrication using an industry 4.0 and XAI approach. The Int. J. Adv. Manuf. Technol. **134**(1), 113–125 (2024)

S. Chiang, Chip giant TSMC opens first Japan factory as it diversifies away from Taiwan amid U.S.-China tensions (2024). https://www.cnbc.com/2024/02/26/tsmc-opens-new-plant-in-japan-as-it-diversifies-away-from-taiwan.html

CommercialEdge, Industrial pipeline slows as deliveries continue to outpace construction starts (2024). https://www.commercialedge.com/blog/national-industrial-report/

D. Dhouib, An extension of MACBETH method for a fuzzy environment to analyze alternatives in reverse logistics for automobile tire wastes. Omega **42**(1), 25–32 (2014)

J. Dong, S. Wan, S.M. Chen, Fuzzy best-worst method based on triangular fuzzy numbers for multi-criteria decision-making. Inf. Sci. **547**, 1080–1104 (2021)

C.A.B. e Costa, J.C. Vansnick, A theoretical framework for measuring attractiveness by a categorical based evaluation technique (MACBETH). in *Multicriteria Analysis* (1997), pp. 15–24

P. Fan, Y. Zhu, Z. Ye, G. Zhang, S. Gu, Q. Shen, E. Alvandi, Identification and prioritization of tourism development strategies using SWOT, QSPM, and AHP: a case study of Changbai Mountain in China. Sustainability **15**(6), 4962 (2023)

M. Farrokhnia, S.K. Banihashem, O. Noroozi, A. Wals, A SWOT analysis of ChatGPT: implications for educational practice and research. Innov. Educ. Teach. Int. **61**(3), 460–474 (2024)

E. Frost, K. Liu, TSMC founder: "In the chip sector, globalization is dead" (2023). https://english.cw.com.tw/article/article.action?id=3396

M. Hanss, *Applied Fuzzy Arithmetic* (Springer, Berlin Heidelberg, 2005)

V.D.B. Huynh, P.V. Nguyen, Q.L.H.T.T. Nguyen, P.T. Nguyen, Application of fuzzy analytical hierarchy process based on geometric mean method to prioritize social capital network indicators. Int. J. Adv. Comput. Sci. Appl. **9**(12), 182–186 (2018)

IRS, Federal income tax rates and brackets (2024). https://www.irs.gov/filing/federal-income-tax-rates-and-brackets

A. Ishizaka, Comparison of fuzzy logic, AHP, FAHP and hybrid fuzzy AHP for new supplier selection and its performance analysis. Int. J. Integrat. Supply Managem. **9**(1–2), 1–22 (2014)

L. James, Why are semiconductor fabs dotting Arizona, the "silicon desert? (2021). https://www.allaboutcircuits.com/news/semiconductor-fabs-dotting-arizona-silicon-desert/

JENKS2026, Navigating regulatory hurdles (2024). https://green.org/2024/01/30/navigating-regulatory-hurdles/

P.-X. Jiang, *Localization Analysis of Semiconductor Supply Chain* (2024). Master Degree Dissertation, National Yang Ming Chiao Tung University.

Y. Kageyama, Taiwan giant chipmaker TSMC opens first plant in Japan as part of key global expansion (2024). https://apnews.com/article/tsmc-semiconductor-chips-taiwan-sony-japan-toyota-2b42cfe27047f68c5edc5228eb27e0c1

Ç.K. Kaymaz, S. Birinci, Y. Kızılkan, Sustainable development goals assessment of Erzurum province with SWOT-AHP analysis. Environ. Dev. Sustain. **24**(3), 2986–3012 (2022)

M.I. Khan, S.G. Al-Ghamdi, Hydrogen economy for sustainable development in GCC countries: a SWOT analysis considering current situation, challenges, and prospects. Int. J. Hydrogen Energy **48**(28), 10315–10344 (2023)

S. Kubler, J. Robert, W. Derigent, A. Voisin, Y. Le Traon, A state-of the-art survey and testbed of fuzzy AHP (FAHP) applications. Expert Syst. Appl. **65**, 398–422 (2016)

Y.C. Lin, T.C.T. Chen, An intelligent system for assisting personalized COVID-19 vaccination location selection: Taiwan as an example. Digital Health **8**, 20552076221109064 (2022)

S. Lundberg, Welcome to the SHAP documentation (2018). https://shap.readthedocs.io/en/latest/index.html

H. Motameni, A. Ebrahimnejad, J. Vahidi, Morphology of composition functions in Persian sentences through a newly proposed classified fuzzy method and center of gravity defuzzification method. J. Intell. Fuzzy Syst. **36**(6), 5463–5473 (2019)

M.Z. Naghadehi, R. Mikaeil, M. Ataei, The application of fuzzy analytic hierarchy process (FAHP) approach to selection of optimum underground mining method for Jajarm Bauxite mine, Iran. Expert Syst. Appl. **36**(4), 8218–8226 (2009)

Y. Nazar, R.A.P. Lovian, D.C. Raharjo, C.N. Rosyidi, Supplier selection and order allocation using TOPSIS and linear programming method at Pt. Sekarlima Surakarta. in AIP Conference Processing, vol. 2097(1), (2019) pp. 030050

H.T.T. Nguyen, H.Q. Cao, K.V.T. Nguyen, N.D.K. Pham, Evaluation of explainable artificial intelligence: Shap, lime, and cam. in *Proceedings of the FPT AI Conference* (2021), pp. 1–6

S. Opricovic, Fuzzy VIKOR with an application to water resources planning. Expert Syst. Appl. **38**(10), 12983–12990 (2011)

B. Phadermrod, R.M. Crowder, G.B. Wills, Importance-performance analysis based SWOT analysis. Int. J. Inf. Manage. **44**, 194–203 (2019)

V. Pillai, Enhancing transparency and understanding in AI decision-making processes. Iconic Res. Eng. J. **8**(1), 168–172 (2024)

PWC, Individual - Taxes on personal income (2024). https://taxsummaries.pwc.com/japan/individual/taxes-on-personal-income

L. Ren, L. Zhang, L. Wang, F. Tao, X. Chai, Cloud manufacturing: key characteristics and applications. Int. J. Comput. Integr. Manuf. **30**(6), 501–515 (2017)

C. Rogers, Everyone loves localization of semiconductors; Applied Materials, Intel act (2021). https://www.spglobal.com/marketintelligence/en/news-insights/latest-news-headlines/everyone-loves-localization-of-semiconductors-applied-materials-intel-act-63661421

K. Roshan, A. Zafar, Using kernel shap xai method to optimize the network anomaly detection model. in *2022 9th International Conference on Computing for Sustainable Global Development* (2022), pp. 74–80

M., Samejima, Y., Shimizu, M., Akiyoshi, N. Komoda, SWOT analysis support tool for verification of business strategy. in *2006 IEEE International Conference on Computational Cybernetics* (2006). pp. 1–4

T.L. Satty, in *The Analytic Hierarchy Process* (McGraw-Hill, 1980)

M. Savari, M.S. Amghani, SWOT-FAHP-TOWS analysis for adaptation strategies development among small-scale farmers in drought conditions. Int. J. Disaster Risk Reduct. **67**, 102695 (2022)

A. Shilov, TSMC completes construction of 5nm fab 21 in Arizona (2022). https://www.tomshardware.com/news/tsmc-fab-21-arizona

H. Shinno, H. Yoshioka, S. Marpaung, S. Hachiga, Quantitative SWOT analysis on global competitiveness of machine tool industry. J. Eng. des. **17**(03), 251–258 (2006)

M. Strong, TSMC sees Taiwan semiconductor business as element of regional stability (2023). https://www.taiwannews.com.tw/news/4911416

The Fund for Peace, Fragile State Index (2024). https://fragilestatesindex.org/

TSMC, JASM set to expand in Kumamoto Japan (2024). https://pr.tsmc.com/english/news/3105

I. Voukkali, A.A. Zorpas, Evaluation of urban metabolism assessment methods through SWOT analysis and analytical hierocracy process. Sci. Total. Environ. **807**, 150700 (2022)

Y.C. Wang, T.C.T. Chen, Analyzing the impact of COVID-19 vaccination requirements on travelers' selection of hotels using a fuzzy multi-criteria decision-making approach. Healthcare Anal. **2**, 100064 (2022)

Y.C. Wang, T. Chen, New XAI tools for selecting suitable 3D printing facilities in ubiquitous manufacturing. Complex and Intell. Syst. **9**(6), 6813–6829 (2023)

Y.-C. Wang, H.-R. Tsai, T. Chen, A selectively fuzzified back propagation network approach for precisely estimating the cycle time range in wafer fabrication. Mathematics **9**(12), 1430 (2021)

Y.C. Wang, T.C.T. Chen, H.C. Wu, A novel auto-weighting deep-learning fuzzy collaborative intelligence approach. Decision Anal. J. **6**, 100186 (2023a)

Y.C. Wang, T.C.T. Chen, Y.C. Lin, 3D printer selection for aircraft component manufacturing using a nonlinear FGM and dependency-considered fuzzy VIKOR approach. Aerospace **10**(7), 591 (2023b)

Y.C. Wang, T. Chen, Adapted techniques of explainable artificial intelligence for explaining genetic algorithms on the example of job scheduling. Expert Syst. Appl. **237**(A), 121369 (2024)

H. Weihrich, The TOWS matrix—A tool for situational analysis. Long Range Plan. **15**(2), 54–66 (1982)

Wikipedia, List of countries by natural disaster risk (2024). https://en.wikipedia.org/wiki/List_of_countries_by_natural_disaster_risk

Wikipedia, State income tax (2024). https://en.wikipedia.org/wiki/State_income_tax

H.-C. Wu, T. Chen, C.-H. Huang, A piecewise linear FGM approach for efficient and accurate FAHP analysis: smart backpack design as an example. Mathematics **8**, 1319 (2020)

H.-C. Wu, T. Chen, S.Y. Wang, Impact of the COVID-19 pandemic on the revenue of the catering industry: Taiwan as an example. SAGE Open **14**(2), 21582440241241410 (2024)

Y. Zhu, C. Chen, G. Zhang, Z. Lin, S.G. Meshram, E. Alvandi, Investigation of West Lake ecotourism capabilities using SWOT and TOPSIS decision-making methods. Sustainability **15**(3), 2464 (2023)

Chapter 5
Rebuilding Semiconductor Manufacturing Competitiveness and Sustainability Through Supply Chain Localization

Abstract This chapter first introduces the concept of competitiveness and identifies the factors that affect a company's competitiveness from a holistic perspective. Subsequently, supply chain competitiveness is defined and the literature on factors affecting supply chain competitiveness is reviewed. The competitiveness of semiconductor manufacturing then comes into focus. Qualitative and quantitative methods for assessing the competitiveness of semiconductor manufacturers are introduced. To illustrate these assessment methods, the impact of recent fab localization initiatives on the competitiveness of these foundries is assessed. Based on the assessment results, the competitiveness of semiconductor manufacturers in various countries are compared. The second part of this chapter introduces the concept of sustainability. A company's short-term competitiveness is often thought to contribute to its long-term sustainability. However, some recent cases have shown that it is possible to pursue long-term sustainability at the expense of short-term competitiveness, such as the recent supply chain localization initiatives of some foundries. Nonetheless, the chapter concludes with a discussion of how to enhance the competitiveness and sustainability of semiconductor manufacturing through localization of supply chains.

Keywords Competitiveness · Sustainability · Wafer fabrication · Semiconductor manufacturing · Localization

5.1 Competitiveness and Influencing Factors

Competitiveness is the ability and performance of a company, subsector or country to sell and supply goods and/or services in a specific market (Chen et al. 2024a). Walsh et al. (2005) claimed that there is a link between competitiveness and the long-term success of an enterprise, highlighting the importance of competitiveness. Porter (1989) derived the five forces that affect corporate competitiveness as **the threat of substitute products**, **the threat of existing competitors**, **the threat of new entrants**, **the bargaining power of suppliers** and **the bargaining power of customers** (see Fig. 5.1). However, the forces are invisible.

T. T. Chen, *Supply Chain Localization in the Semiconductor Industry*,
https://doi.org/10.1007/978-3-031-81280-4_5

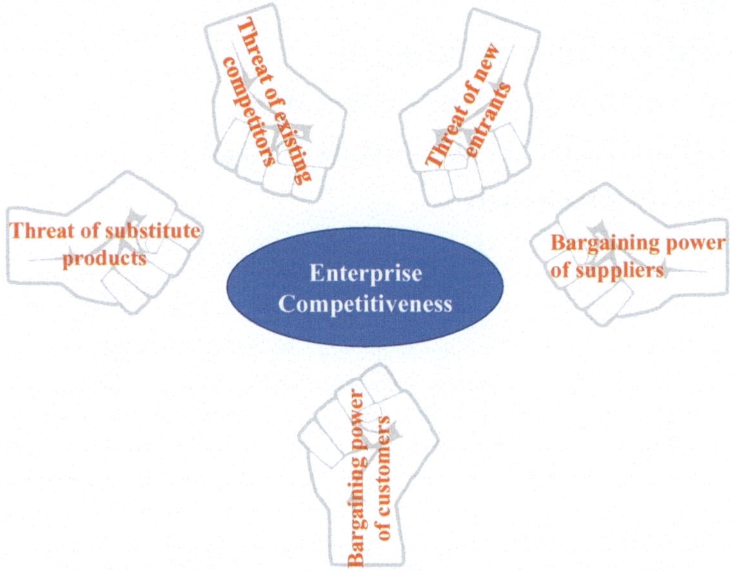

Fig. 5.1 Five forces that affect corporate competitiveness

Ivanova et al. (2018) evaluated the competitiveness of a company by taking into account the three major types of economic activity: supply, production and sales. The poor performance in any one of these economic activities can prevent a company from competing with its competitors.

Golovchenko et al. (2022) asserted that competitiveness is the only criterion for the efficiency and effectiveness of an economic entity in the era of globalization and internationalization of the economy.

If a company is highly competitive, according to Ivanova et al. (2018), it will have a stable and cheap supply, maximize its output, and obtain large sales. As a result, the company earns more profits and gains higher market share. Therefore, according to Golovchenko et al. (2022), such a company operates both efficiently and effectively. This viewpoint evaluates efficiency and effectiveness from the inside out (see Fig. 5.2).

In Fig. 5.3, the five forces derived by Porter (1989) are considered to have an influence on the three goals in Fig. 5.3, showing that these five forces can drag a company down from achieving the three goals.

Most of the literature in the field of competitiveness focus on factors that affect competitiveness (such as cost, quality, customer satisfaction, technical capabilities, etc.) and methods to improve competitiveness. Although competitiveness has been recognized as a key issue, there is no consensus as to what quantitatively constitutes a high or low level of competitiveness (Chen 2023a).

In the opinion of Shinno et al. (2006), the global competitiveness of an enterprise includes three parts:

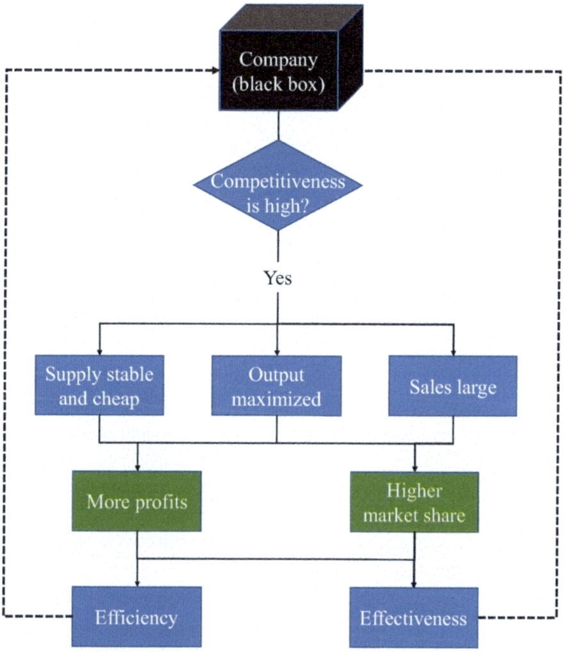

Fig. 5.2 Competitiveness as the only criterion for the efficiency and effectiveness of a company

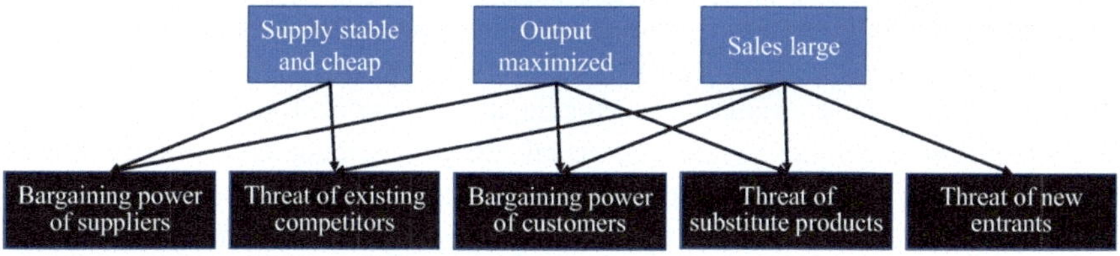

Fig. 5.3 Influence of the five forces derived by Porter on the three aspects in Fig. 5.2

- Market-related competitiveness;
- Organization-related competitiveness;
- Product-related competitiveness.

Factors affecting the three parts of the global competitiveness are summarized in Table 5.1.

5.2 Supply Chain Competitiveness

With the trend of globalization, the competition within some industries has become increasingly fierce. In order to survive in the industry, every company must strive to continuously improve its capability in some way (Chen 2023b). For example,

Table 5.1 Factors affecting the three parts of the global competitiveness

Part of the Global Competitiveness	Affecting Factors
Market-related competitiveness	• Adaptability to customer requirements • Multi-faceted support from government policies • Development of new market • Leading position in production volume • Favorable market competition environment
Organization-related competitiveness;	• On-time delivery • Short product development cycle • Human resource development • Possible industrial restructuring • Improvement of profitability • Strengthening the cooperation between industrial and academic organizations
Product-related competitiveness	• Product quality • Introduction of advanced technology • Supplier participation in product development • Total quality of components and parts • Introduction of new materials

some companies (such as chip designers) do not have their own factories so they can focus on more profitable activities, while others (such as wafer foundries) continue to expand fabrication capabilities to further reduce unit costs. Other common strategies include: B2B e-commerce, outsourcing, blue ocean strategy, better scheduling, factory simulation, green and lean technologies, competitiveness diamond model, cyber-physical systems, cloud manufacturing, developing next-generation technologies, forming virtual alliances (such as virtual enterprises, financial networks, service networks, supply and demand chains, etc.) and others.

Virtual alliances are highly efficient and competitive. Chen et al. (2023a) mentioned the three issues emerged in this field:

- **Green competitiveness**: Green competitiveness aims to adopt various strategies to achieve high productivity and overall socio-economic development while maintaining the goal of sustainable development.
- **Competitiveness versus productivity**: Since competitiveness can lead to wasteful expenditures, productivity is crucial in a factory.
- **Sustainable competitiveness**: Companies move to regions with lower wages to maintain competitive advantages.

The last issue is compounded by **supply chain localization**, as suppliers must move closer to customers, potentially at the expense of higher wages.

Supply chains are a special type of virtual systems. The competitiveness of a supply chain can be defined similarly to that of an enterprise (Chen et al. 2024a) as **the ability and performance of a supply chain to sell and supply goods and/or services in a specific market**. However, Verma and Seth (2010) believed that supply chain competitiveness is different from enterprise competitiveness. The difference is that to be competitive, a supply chain must be able to create and deliver value to its participants (including upstream and downstream suppliers), not just its customers.

Joshi et al. (2013) tried to identify the factors affecting the competitiveness of India's auto parts supply chains and thus established twenty criteria to evaluate the competitiveness of a supply chain. These criteria were divided into seven categories: **costs**, **quality**, **technologies**, **environmental factors**, **heterogeneous factors**, **buyer–supplier relationships** and **flexibility**. According to their research, business environment factors such as worker skills, globalization and government regulations contributed the most to the competitiveness of India's auto parts supply chains.

Febransyah and Camelia Goni (2022) considered four criteria for assessing supply chain competitiveness: **costs**, **differentiation**, **sustainability**, and **infrastructure**. They then applied analytical hierarchy process (AHP), a well-known multi-criteria decision-making (MCDM) method, to fulfill the assessment task. The experimental results showed that in order to gain supply chain competitiveness, companies must implement strategies focused on performances such as costs, differentiation and sustainability, as well as strategies on internal infrastructure such as software and hardware.

However, most of the past research on supply chain competitiveness focused on how to improve, rather than how to measure, supply chain competitiveness. For example, according to Pine (1993), the competitiveness of a supply chain depends on the quantity of products made through the system. **Mass customization** is therefore critical to gain supply chain competitiveness by significantly increasing variety and customization without sacrificing efficiency, effectiveness or costs.

O'Connor et al. (2018) summarized the five main challenges in gaining brand competitiveness in China's supply chains as **costs**, **competition**, **sales growth**, **product development** and **human resources**. To overcome these challenges, supply chains developed strategies of market competition, sales growth, international customers and strategic market fit.

The lasting competitiveness of an enterprise (or supply chain) brings about its sustainability. Tseng et al. (2021) fitted the relationship between supply chain competitiveness and a set of structural attributes as well as the causal relationship between them. To cope with the fuzziness of human decision-making and the complexity and uncertainty of the supply chain (Chen et al. 2024b, 2024c), the fuzzy Delphi method and fuzzy decision making and trial evaluation laboratory (DEMATEL) were applied. The research results showed that **digital platform effectiveness**, **digital communication** from the perspective of supply chain digitization, **labor conditions** and **manufacturing processes** were important for sustainable supply chain management. In addition, the most important cause-and-effect criteria, including logistics integration, logistics optimization, delivery speed, proactive actions and just-in-time inventory, were considered useful in helping supply chain participants make better practices.

Fig. 5.4 Factors contributing to supply chain competitiveness

Sarker et al. (2022) conducted an extensive literature review on supply chain competitiveness and identified six key pillars of supply chain competitiveness: **innovation**, **sustainability**, **collaboration**, **information technology**, **agility**, and **flexibility**. Furthermore, they asserted that for supply chains to remain competitive in the post-COVID-19 era, the combination of these pillars, if not all, are required.

Factors contributing to supply chain competitiveness are summarized in Fig. 5.4.

5.3 Semiconductor Manufacturing Competitiveness

The competition within the semiconductor industry is fierce (Chen 2023c). For example, according to the statistics from Wikipedia (2024), 45 of the 530 wafer fabs have been closed due to their low competitiveness.

The competition within the semiconductor industry has led to the continued development of larger circuits that use more transistors per unit area while further reducing sizes without sacrificing the performances of outdated technologies (Mcfee 2023). For example, wafer foundries are currently developing metal oxide semiconductors to further improve efficiency while reducing sizes more significantly than before, so as to control power consumption levels as consumer electronics devices become more commonplace in daily life activities. Additionally, facing stiff competition from rivals, chip designers must strive to produce new designs that offer superior performance at lower prices (Chen 2023d). This competition is critical to driving innovation in the semiconductor industry and lowering costs for buyers (Wang and Chen 2024;

Chen 2023e). It also ensures consumers enjoy a wide range of cutting-edge products at the best prices (Mcfee 2023).

In order to survive in such a highly competitive industry, semiconductor manufacturers enhance their competitiveness in a variety of ways. For example, Armstrong (1989) proposed four principles for competitive semiconductor manufacturing: **continuous measurable improvement**, **statistical thinking**, **constraint focus**, and **people development/empowerment**. Both Jenkins et al. (1990) and Crowder (1989) emphasized the importance of **quality** and introduced some methods to design quality into products and processes. Peng and Chien (2003) focused on how to create value for customers. **Shortening cycle time** (Wang et al. 2023a; Udokporo et al. 2020; Chen 2023f), **producing high-quality products**, **delivering orders on time** (Papageorgiou et al. 2020; Chen 2023g), **continuously reducing costs**, and **improving efficiency** were considered to be the most direct and effective methods. Liao and Hu (2007) pointed out that **knowledge transfer** is an important factor in the competitiveness of semiconductor manufacturers. On the contrary, according to Crowder (1989), the leadership in manufacturing sciences and technologies does not necessarily lead to an competitive advantage.

Until now, most semiconductor industry leaders have derived their competitiveness from government support. The leaders also urged their governments to erect barriers to semiconductors from other countries. From this point of view, Khan et al. (2021) evaluated and compared the national competitiveness of different countries. They clearly demonstrated the advantages of the United States and its allies.

Singh et al. (2020) listed the sources for the competitiveness of Taiwan Semiconductor Manufacturing Company (TSMC), the largest wafer foundry in the world, as

- The time and cost of building a new wafer fab are unaffordable for new competitors.
- It is difficult for new competitors to recruit sufficient skilled labor.
- The price of advanced wafer fabrication equipment is increasing day by day, and it takes a long time to obtain such equipment, which is detrimental to new competitors.
- TSMC enjoys advanced packaging patents.

5.3.1 Qualitative Assessment Methods

Being in the right market at the right time and producing the right products seems to be a prerequisite for companies to remain highly competitive. Chen and Wang (2024a) used a tree diagram (see Fig. 5.5) to summarize the factors that affect a company's competitiveness and their interactions. Such factors include capacity, technologies, employee quality, costs (including budgets), cycle time, yields (product quality), work in process (WIP), prices, flexibility and responsiveness. The interactions of these factors can lead to higher profits, higher customer satisfaction, and higher market share.

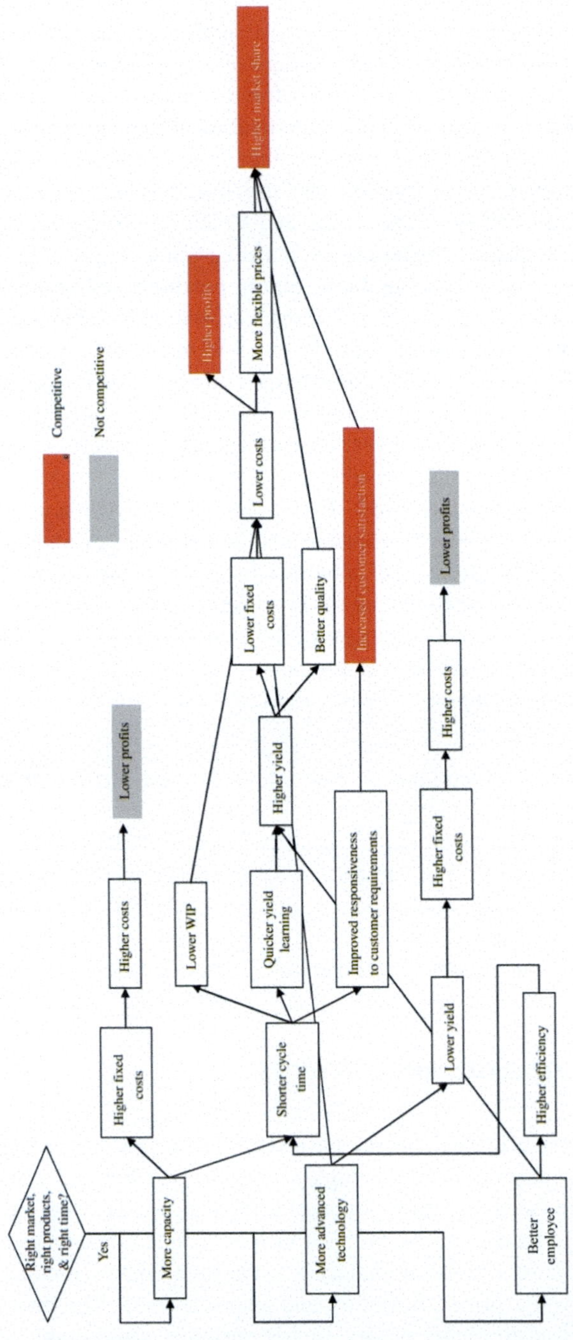

Fig. 5.5 Factors affecting a company's competitiveness

Currently, political factors (geopolitics (Chiang 2024) or the US-China trade war (Bown 2020)) are the primary factors affecting the competitiveness of semiconductor manufacturers. Chip designers choose wafer foundries that meet political requirements. Secondly, that also depends on the market being targeted. For semiconductor manufacturers providing products for mature markets, costs may be the most critical (Chen 2023h) (see Fig. 5.6), while for semiconductor manufacturers providing advanced products for emerging markets, technologies are crucial (Tamirat 2017) (see Fig. 5.7), in which yellow and green lines indicate satisfaction and the opposite statuses, respectively.

Furthermore, recent research on semiconductor manufacturing competitiveness has mainly focused on differentiating the competitiveness of countries and how to improve the competitiveness of the semiconductor industry in a certain country (Grimes and Du 2022; Lee 2023), which is usually the responsibility of the country's government. The localization of semiconductor supply chains is also driven in part by the need for governments to upgrade their domestic semiconductor manufacturing industries. Due to the large amount of investment involved, a successful semiconductor supply chain localization plan will also become a political achievement for the local government, making political factors more complicated. Nonetheless, the same tree diagram can be used to evaluate the impact of a wafer fab localization plan, such as that of TSMC in Arizona (Vandendriessche and Buts 2024), on its competitiveness (see Fig. 5.8). TSMC is the largest wafer foundry in the world.

5.3.2 Quantitative Assessment Methods: Analytic Hierarchy Process (AHP) Method

Shinno et al. (2006) defined the three parts of the global competitiveness of an enterprise as market-related competitiveness, organization-related competitiveness, and product-related competitiveness. They then listed factors affecting the three parts of the global competitiveness, based on which the analytic hierarchy process (AHP) method was applied to evaluate the competitiveness of an enterprise. A special feature of their approach was that the affecting factors were divided into four categories of the enterprise, following the concept of strength, weakness, opportunity and threat (SWOT) analysis.

The decision-making hierarchy is shown in Fig. 5.9. In addition, factors at the lowest level can be further decomposed, as shown in Table 5.1. To solve the AHP problem, eight sub-problems need to be addressed. The results are then aggregated, as illustrated by Fig. 5.10. In this figure, the weight (priority) of strengths to the global competitiveness can be derived as

$$\text{Strengths} : 0.45^{*}0.26 + 0.25^{*}0.17 + 0.3^{*}0.22 = 0.2255.$$

Similarly, the weights of weakness, opportunities and threats can be obtained as

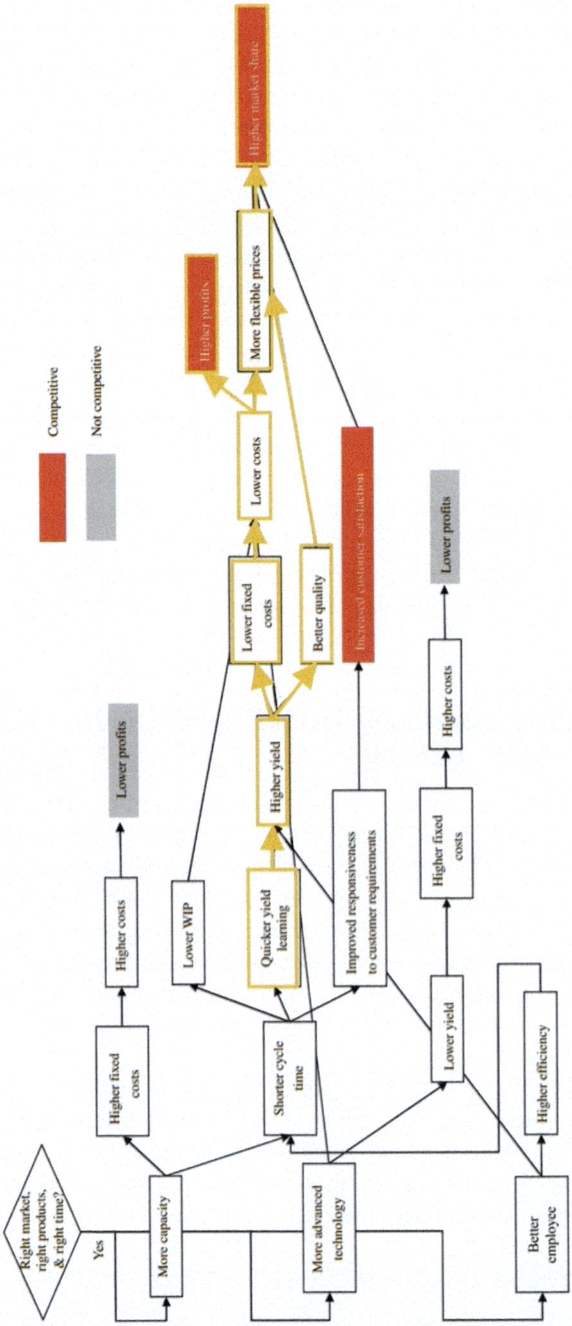

Fig. 5.6 Factors affecting the competitiveness of semiconductor manufacturers providing products for mature markets

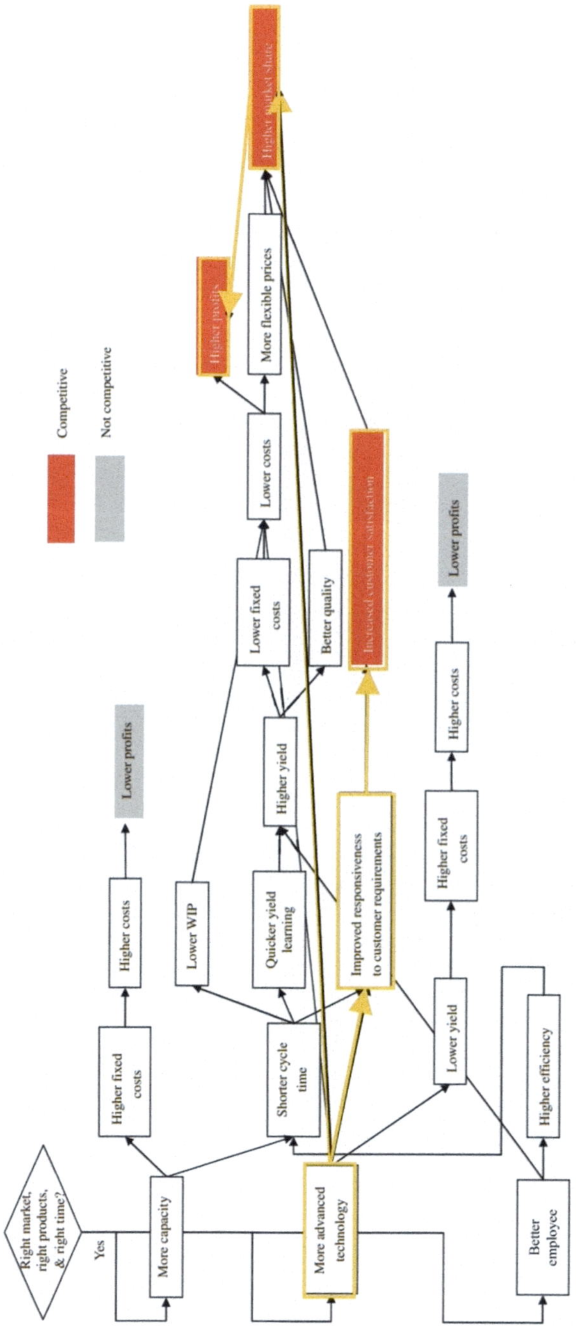

Fig. 5.7 Factors affecting the competitiveness of semiconductor manufacturers providing products for emerging markets

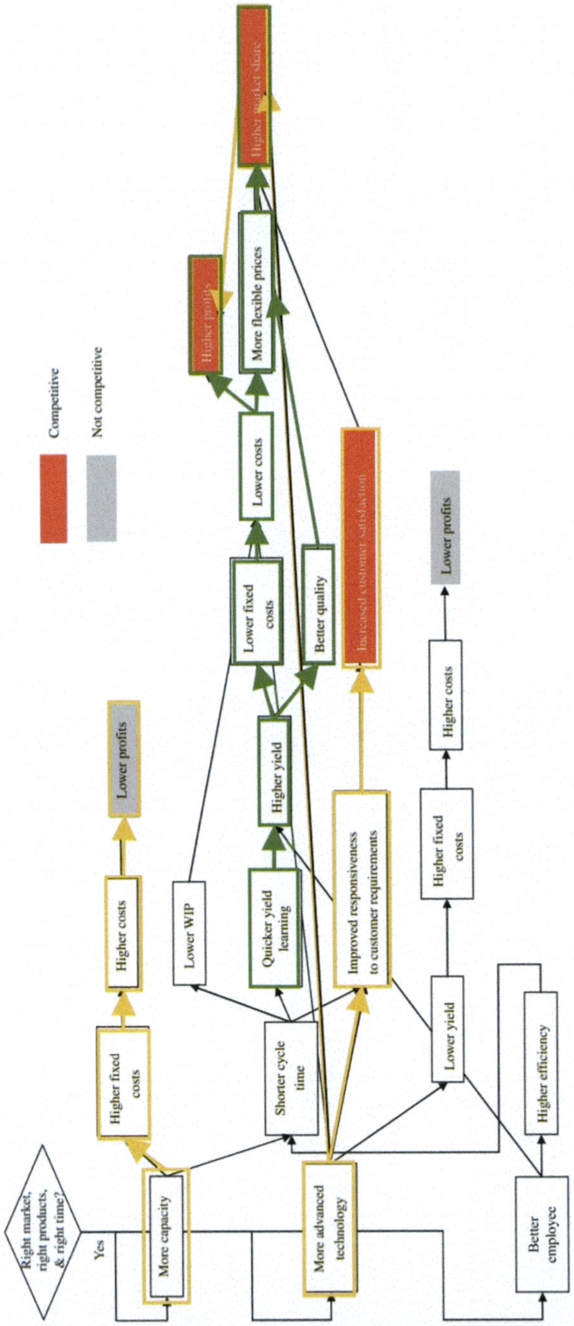

Fig. 5.8 Evaluation of the impact of TSMC's wafer fab localization initiative in Arizona, USA on its competitiveness

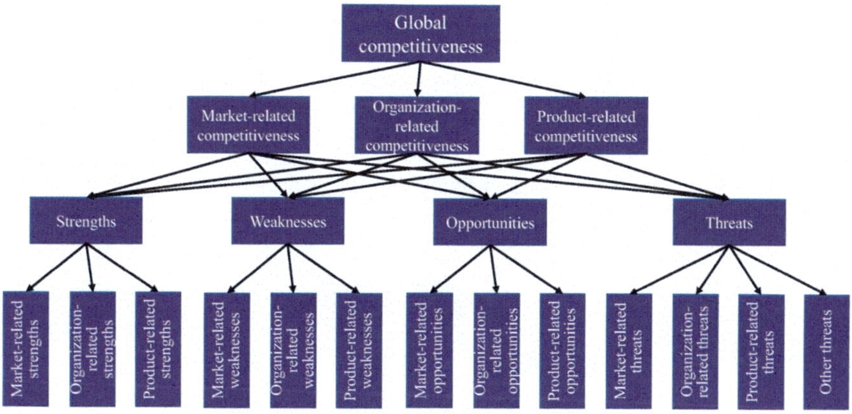

Fig. 5.9 Decision-making hierarchy

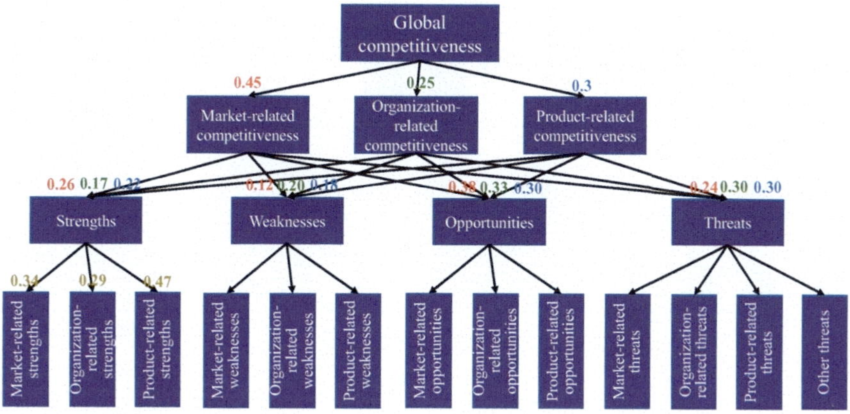

Fig. 5.10 Aggregation

$$\text{Weakness} : 0.143.$$

$$\text{Opportunities} : 0.3435.$$

$$\text{Threats} : 0.273.$$

The weight of market-related strengths to the global competitiveness at the top can then be derived as

$$\text{Market} - \text{related strengths} : 0.2255^{*}0.34 = 0.07667.$$

After collecting the performances of the factors at the lowest level, and multiplying the collected performances to the weights of these factors, the global competitiveness can be evaluated. An example is provided in Fig. 5.11. The global competitiveness of the enterprise can be evaluated as (Wu et al. 2024; Chen and Lee 2024a).

The global competitiveness = 0.0767 * 4 + 0.0654 * 3.5 + 0.1060 * 2.5 + 0.0544 * 4.5 + 0.0279 * 5 + 0.0607 * 3 + 0.1225 * 3.5 + 0.1309 * 2 + 0.0901 * 2.5 + 0.0767 * 4 + 0.0885 * 4 + 0.0721 * 3.5 + 0.1124 * 2 = 3.42.

Such global competitiveness can be said to be high (Wang et al. 2023b) (see Fig. 5.12).

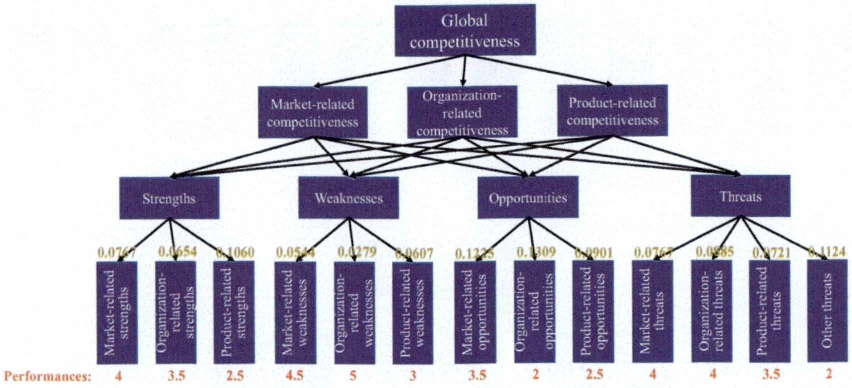

Fig. 5.11 Global competitiveness evaluation

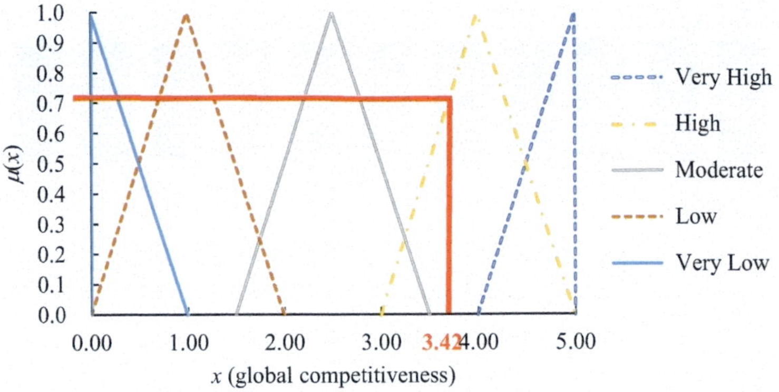

Fig. 5.12 Determining the global competitiveness

5.3.3 *Quantitative Assessment Methods: Fuzzy Yield Learning Method*

Yield is undoubtedly the most critical factor in the competitiveness of semiconductor manufacturers. Therefore, it is a reasonable idea to use yield to evaluate the competitiveness of a semiconductor product. In addition, product unit costs, which are critical to fab competitiveness, also depend on their yields, further reinforcing the incentive for adopting the yield-based competitiveness. Then, the overall competitiveness of a semiconductor manufacturer can be assessed from those of all its products.

The fuzzy yield learning method for evaluating the competitiveness of a semiconductor product comprises the following steps:

- Select a suitable (fuzzy) yield learning model.
- Fit the parameter values in the (fuzzy) yield learning model.
- Set up a yield target.
- Establish the competitive region.
- Evaluate the mid-term competitiveness.
- Take actions to enhance the mid-term competitiveness if it is negative.

as illustrated in Fig. 5.13.

The mid-term competitiveness of a semiconductor product is evaluated based on its yield learning model:

$$Y_t = Y_0 e^{-\frac{b}{t} + r(t)} \tag{5.1}$$

or in logarithms,

$$\ln(Y_t) = \ln(Y_0) - \frac{b}{t} + r(t) \tag{5.2}$$

where Y_t is the yield forecast at time period t; Y_0 is the asymptotic/final yield, which is a real-valued function of point defect density per unit area, chip area, and a set of parameters unique to the specific yield model; b is the learning constant; $r(t)$ is a homoscedastical, serially non-correlated error term:

$$r(t) \sim \text{Normal}(0, \sigma^2),\ \hat{r}(t) = 0 \quad \text{for all } t \tag{5.3}$$

where "^" denotes that it is an estimate. We can obtain

$$\hat{b} = -\frac{S_{xy}}{S_{xx}} \tag{5.4}$$

$$\hat{Y}_0 = e^{\frac{1}{T}\sum_{t=1}^{T}\ln(Y_t) + \frac{\hat{b}}{T}\sum_{t=1}^{T}\frac{1}{t}} \tag{5.5}$$

where

Fig. 5.13 Steps of the fuzzy yield learning method

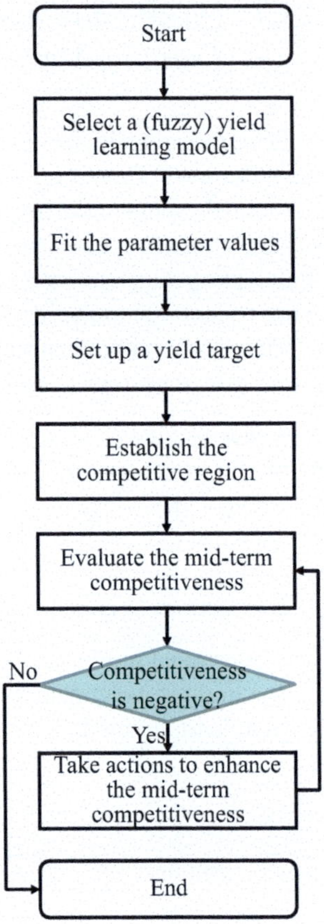

$$S_{xx} = \sum_{t=1}^{T} \frac{1}{t^2} - \frac{(\sum_{t=1}^{T} \frac{1}{t})^2}{T} \tag{5.6}$$

$$S_{xy} = \sum_{t=1}^{T} \frac{\ln(Y_t)}{t} - \frac{1}{T}(\sum_{t=1}^{T} \frac{1}{t})(\sum_{t=1}^{T} \ln(Y_t)) \tag{5.7}$$

The fitness of this model can be evaluated by the coefficient of determination (R^2):

$$R^2 = \frac{S_{xy}^2}{S_{xx}S_{yy}} \quad \text{(from 0 (the worst) to 1 (the best))} \tag{5.8}$$

where

$$S_{yy} = \sum_{t=1}^{T} \ln(Y_t)^2 - \frac{(\sum_{t=1}^{T} \ln(Y_t))^2}{T} \tag{5.9}$$

Subsequently, the mid-term competitiveness of the product is evaluated based on the yield learning model as follows. Assume the competitive region is specified as

$$Y_t \geq Y_m^* \quad \text{when } t \geq m \tag{5.10}$$

Then the mid-term competitiveness is evaluated with the area of the competitive region that is under the yield learning curve from time m to $m + \Delta t$:

$$\begin{aligned} \text{Competitiveness} &= \int_m^{m+\Delta t} Y_t dt - \int_m^{m+\Delta t} Y_m^* dt \\ &= \int_m^{m+\Delta t} \hat{Y}_0 e^{-\frac{\hat{b}}{t}} dt - Y_m^* \Delta t \end{aligned} \tag{5.11}$$

where Δt is a small value (≤ 1) indicating the time allowance for achieving the mid-term yield target. If $\Delta t = 1$, then the calculation result ranges from $-Y_m^*$ to $1-Y_m^*$. If the result is negative, then the product will not have mid-term competitiveness.

Example 5.1 The example in Table 5.2 (Chen 2023c) is analyzed. At first, the fitted yield learning model is

$$Y_t = 0.77e^{-\frac{0.7}{t}+r(t)}$$

Assume the competitive region is specified as

Table 5.2 Yield data of a semiconductor product

t	Y_t (%)
1	37
2	59
3	54
4	74
5	62
6	80
7	72
8	72
9	59
10	72

$$Y_t \geq 76\% \text{ when } t \geq 15;\ \Delta t = 1$$

So $m = 15$ and $Y_m^* = 76\%$. Then the mid-term competitiveness of the semiconductor product can be evaluated according to Eq. (5.11) as

$$\text{Competitiveness} = \int_{15}^{15+1} 0.77e^{-\frac{0.7}{t}} dt - 0.76 \cdot 1 = -0.02$$

Since the evaluation result is negative, it can be concluded that the semiconductor will not have competitiveness at time period 15, unless some managerial actions (e.g. increasing the speed of mass production or executing quality engineering (QE) projects) are taken to quicken the yield learning process (Chen 2023i).

However, there are many uncertainties in a yield learning process. To address this issue, Chen (2023c) established a systematic procedure based on the fuzzy yield learning model of a semiconductor to evaluate its mid-term yield competitiveness:

$$\tilde{Y}_t = \tilde{Y}_0 e^{-\frac{\tilde{b}}{t}+r(t)} \tag{5.12}$$

After lognormalization:

$$\ln(\tilde{Y}_t) = \ln(\tilde{Y}_0)(-)\frac{\tilde{b}}{t} + r(t) = \left(\ln Y_{01} - \frac{b_3}{t}, \ln Y_{02} - \frac{b_2}{t}, \ln Y_{03} - \frac{b_1}{t}\right) + r(t) \tag{5.13}$$

where (−) represents fuzzy subtraction. This is a fuzzy linear regression (FLR) problem. The parameters can be obtained by solving the following linear programming (LP) model:

$$\text{Min} \sum_{t=1}^{T} (\ln Y_{03} - \frac{b_1}{t} - (\ln Y_{01} - \frac{b_3}{t})) \tag{5.14}$$

s.t.

$$\ln Y_t \geq \ln Y_{01} - \frac{b_3}{t} + \alpha(\ln Y_{02} - \frac{b_2}{t} - \ln Y_{01} + \frac{b_3}{t}) \tag{5.15}$$

$$\ln Y_t \leq \ln Y_{03} - \frac{b_1}{t} + \alpha(\ln Y_{02} - \frac{b_2}{t} - \ln Y_{03} + \frac{b_1}{t}) \tag{5.16}$$

$$0 \leq b_1 \leq b_2 \leq b_3 \tag{5.17}$$

$$\ln Y_{01} \leq \ln Y_{02} \leq \ln Y_{03} \leq 0 \tag{5.18}$$

$$0 \leq \alpha \leq 1 \tag{5.19}$$

Then the fuzzy mid-term competitiveness can be evaluated as

$$\begin{aligned} \text{Competitiveness} &= \int_{m}^{m+\Delta t} \tilde{Y}_t dt - \int_{m}^{m+\Delta t} Y_m^* dt \\ &= \left(\int_{m}^{m+\Delta t} \hat{Y}_{01} e^{-\frac{\hat{b}_3}{t}} dt, \int_{m}^{m+\Delta t} \hat{Y}_{02} e^{-\frac{\hat{b}_2}{t}} dt, \int_{m}^{m+\Delta t} \hat{Y}_{03} e^{-\frac{\hat{b}_1}{t}} dt \right) - Y_m^* \Delta t \end{aligned} \tag{5.20}$$

In addition, new correction functions were also designed to incorporate expert opinions on medium-term yield targets into the fuzzy yield learning model to maintain the yield competitiveness. These expert opinions were valuable for controlling the yield learning process. The first correction function, $SC(n, t, \tilde{Y}_0, \tilde{Y}_0^*)$, is designed to incorporate expert opinions about the asymptotic yield in the learning model

$$SC(n, t, \tilde{Y}_0, \tilde{Y}_0^*) = e^{(1-\frac{n}{t}) \ln \frac{\tilde{Y}_0^*}{\tilde{Y}_0}} \tag{5.21}$$

where n is the current period; $\tilde{Y}_0^*$ indicates the asymptotic yield suggested by experts, which is a subjective value and expressed in fuzzy numbers or linguistic terms defined in advance.

Chen and Wang's fuzzy yield learning model after correction becomes

$$\begin{aligned} \tilde{Y}_t' &= SC(n, t, \tilde{Y}_0, \tilde{Y}_0^*) \tilde{Y}_0 e^{-\frac{\tilde{b}}{t} + r(t)} \\ &= e^{(1-\frac{n}{t}) \ln \frac{\tilde{Y}_0^*}{\tilde{Y}_0}} (\times) \tilde{Y}_0 e^{-\frac{\tilde{b}}{t} + r(t)} \\ &= \tilde{Y}_0^* e^{-\frac{1}{t}(n \ln \frac{\tilde{Y}_0^*}{\tilde{Y}_0} + \tilde{b})} + r(t) \end{aligned} \tag{5.22}$$

After evaluation, if a semiconductor product is not competitive, then the semiconductor manufacturer must improve the competitiveness of the semiconductor product. Otherwise, the mass production of this semiconductor product will be gradually stopped. To avoid this situation, part of the fabrication capacity for other semiconductor products needs to be withdrawn and reallocated to this semiconductor product to speed up its mass production, thereby accelerating the yield learning process of the semiconductor product. The impact of capacity reallocation on the competitiveness of the semiconductor product is analyzed as follows.

According to Chen and Wang's fuzzy yield learning model, the time to achieve the mid-term yield target Y_m^*, $\tilde{t}^*$, satisfies

$$Y_m^* = \tilde{Y}_0 e^{-\frac{\tilde{b}}{\tilde{t}^*}+r(\tilde{t}^*)} \tag{5.23}$$

So we obtain

$$\tilde{t}^* \cong \frac{\tilde{b}}{\max(\ln \tilde{Y}_0 - \ln Y_m^*,\ 0)} \tag{5.24}$$

If $\tilde{t}^* \leq m$, then the mid-term yield target can be achieved before the required deadline, and the semiconductor product will be competitive. Otherwise, some capacity has to be re-allocated to the semiconductor product to accelerate the yield learning process. Assume originally the capacity allocated for making the semiconductor product is to release q wafer per period. Until time $\tilde{t}^*$, the number of wafers that have been released is

$$q\tilde{t}^* = \frac{q\tilde{b}}{\max(\ln \tilde{Y}_0 - \ln Y_m^*,\ 0)} \tag{5.25}$$

So the number of wafers that still need to be released before time m to achieve the mid-term yield target is

$$\tilde{q}_r = \max\left(\frac{q\tilde{b}}{\max(\ln \tilde{Y}_0 - \ln Y_m^*,\ 0)} - qn,\ 0\right) \tag{5.26}$$

where n denotes the current period. Divide $\tilde{q}_r$ by the available time m–n, the capacity required for the semiconductor product to become mid-term competitive can be obtained as

$$\tilde{q}' = \max\left(\frac{q\tilde{b}}{\max(\ln \tilde{Y}_0 - \ln Y_m^*,\ 0)(m-n)} - \frac{qn}{(m-n)},\ 0\right) \tag{5.27}$$

After subtracting the original capacity, the additional capacity required can be obtained as follows:

$$\Delta\tilde{q} = \tilde{q}' - q = \max\left(\max\left(\frac{q\tilde{b}}{\max(\ln \tilde{Y}_0 - \ln Y_m^*,\ 0)(m-n)} - \frac{qn}{(m-n)},\ 0\right) - q,\ 0\right) \tag{5.28}$$

If $\Delta\tilde{q} = \infty$, then it is impossible to enhance the competitiveness of the semiconductor product by just allocating more capacity.

Example 5.2 In the previous example, assume originally the capacity for the semiconductor product is about 5000 wafers per month, i.e., $q = 5000$. The monthly capacity of the wafer fab is 30000. To enhance the competitiveness of the semiconductor product, the additional capacity required for the semiconductor product is calculated according to Eq. (5.28) as

$$
\begin{aligned}
\Delta q &= q' - q \\
&= \max(\max(\frac{5000 \cdot 0.7}{\max(\ln 0.77 - \ln 0.76,\ 0)(15 - 10)} - \frac{5000 \cdot 10}{(15 - 10)},\ 0) - 5000,\ 0) \\
&= 38549
\end{aligned}
$$

The result indicates that an additional capacity of up to 38,549 wafers per month is needed to enhance the competitiveness of the semiconductor product. This is far more than the fab's monthly fabrication capacity. As a result, it is not possible that the product may turn competitive even if it occupies all fab capacity. Conversely, reallocating capacity will not make any good on the competitiveness of the product if it is already competitive.

Semiconductor manufacturing factories typically have many products. If all products are competitive in the medium term, then there is no doubt that factories will be competitive as well. On the other hand, if certain products are not competitive, the mid-term competitiveness of the factory cannot be guaranteed, especially when the demand for such products is high in the future. To prevent this from happening, capacity can be reallocated to less competitive products, with the overall goal of increasing the competitiveness of the entire factory rather than focusing on the competitiveness of a single product.

Assume there are K products manufactured in the factory. The final yield and learning constant of product k are denoted with $\tilde{Y}_{0(k)}$ and $\tilde{b}_{(k)}$, respectively; $k = 1 \sim K$. The mid-term yield target for product k to achieve is denoted with $(m_{(k)}, Y^*_{m(k)})$. The competitiveness of product k, $\tilde{c}_{(k)}$, can be evaluated according to Eq. (5.20) as

$$
\tilde{c}_{(k)} = \tilde{Y}_{0(k)} \int_{m_{(k)}}^{m_{(k)}+\Delta t} e^{-\frac{\tilde{b}_{(k)}}{t}} dt - Y^*_{m(k)} \Delta t \tag{5.29}
$$

Besides, assume the capacity originally allocated for making product k is $q_{(k)}$. The capacity of the whole factory is Q. Therefore,

$$
\sum_{k=1}^{K} q_{(k)} = Q \tag{5.30}
$$

After capacity reallocation, the capacity for product k becomes $q'_{(k)}$, and the evaluated competitiveness changes to $\tilde{c}'_{(k)}$. According to Eq. (5.29),

$$
\tilde{c}'_{(k)} = \tilde{Y}_{0(k)} \int_{m_{(k)}}^{m_{(k)}+\Delta t} e^{-\frac{\tilde{b}_{(k)} q_{(k)}}{q'_{(k)} t}} dt - Y^*_{m(k)} \Delta t \tag{5.31}
$$

Also,

$$\sum_{k=1}^{K} q'_{(k)} = Q \tag{5.32}$$

The competitiveness of the semiconductor manufacturing factory can be measured in several ways, e.g. the average competitiveness, the minimal competitiveness, the weighted competitiveness, etc. Besides, a small-scale capacity reallocation plan causing little disturbance is also favored. Taking these into consideration, the following fuzzy nonlinear programming (FNP) model (Chen 2023j; Wang and Chen 2023) model is formulated to maximize the competitiveness of the semiconductor manufacturing factory:

$$\text{Max } 1/K \sum_{k=1}^{K} \tilde{c}'_{(k)} \tag{5.33}$$

s.t.

$$\tilde{c}'_{(k)} = \tilde{Y}_{0(k)} \int_{m_{(k)}}^{m_{(k)}+\Delta t} e^{-\frac{\tilde{b}_{(k)} q_{(k)}}{q'_{(k)} t}} dt - Y^{*}_{m(k)} \Delta t \tag{5.34}$$

$$\sum_{k=1}^{K} q'_{(k)} = Q \tag{5.35}$$

$$\sum_{k=1}^{K} \left| q_{(k)} - q'_{(k)} \right| \le \omega \tag{5.36}$$

$$q'_{(k)} \in \mathbf{Z}^{+}, k = 1 \sim K \tag{5.37}$$

The FNP model can be converted into an equivalent nonlinear programming (NP) problem (Chen and Wang 2022a) that can be easily solved with existing optimization software.

While localization attempts by wafer foundries appear to be detrimental to their competitiveness, as localization in building localized wafer fabs is forced rather than optimized, these attempts will not significantly change the competitiveness of wafer foundries, because the fabrication capacity of a localized wafer fab is only a small part of the wafer foundry's total capacity, and therefore customers will not insist that all their products have to be fabricated by localized wafer fabs.

5.3.4 *Competitiveness of Semiconductor Manufacturers in Various Countries*

Taiwan and South Korea are two countries with very strong semiconductor industries, especially wafer fabrication and assembly, test and packaging (ATP). For example, TSMC in Taiwan is the world's largest wafer foundry, and its high competitiveness comes from many sources. South Korea accounts for 60.5% of the global storage semiconductor market and continues to focus on R&D and investment to maintain a competitive advantage. In addition, South Korea also accounts for 17.3% of the global foundry market (Invest Korea2023).

In Taiwan, national policies assist in the development of the semiconductor industry (see Fig. 5.14):

- **Government-established R&D institutions**: In the 1970s, Industrial Technology Research Institute (ITRI) established the Electronics Institute, and with government assistance, ITRI introduced relevant integrated circuit (IC) manufacturing technologies and talents from the United States to cultivate domestic semiconductor talents.

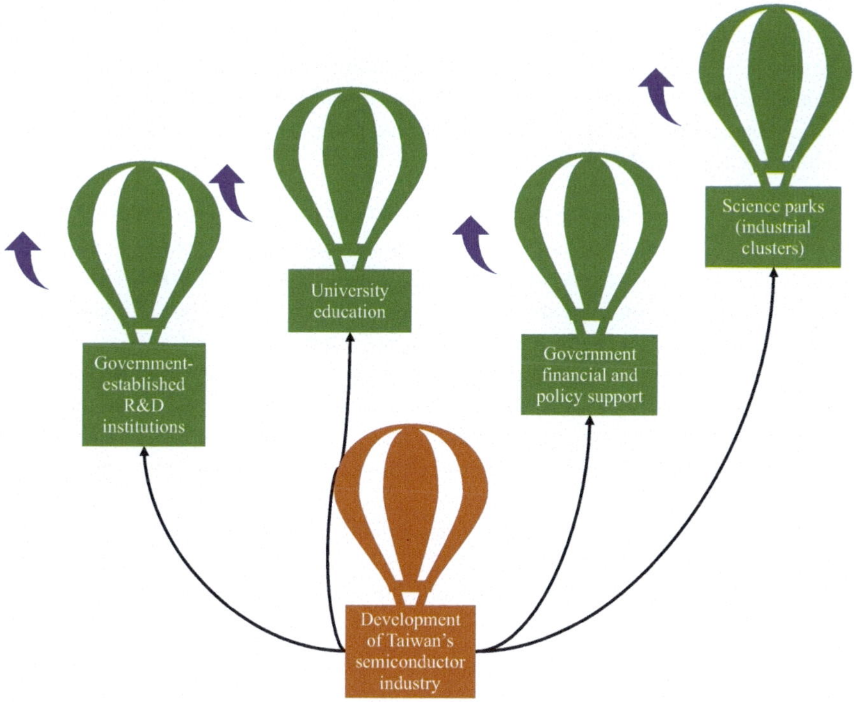

Fig. 5.14 National policies that assist in the development of Taiwan's semiconductor industry

- **University education**: In addition, some universities have established related departments to teach semiconductor manufacturing related knowledge. Human resource costs and the number of technical talents are important factors in assessing the quality of a region's population, which affect the quality and efficiency of wafer fabrication, and indirectly influence the stability and smoothness of semiconductor supply chain operations.
- **Government financial and policy support**: Government support is definitely the most effective means to cultivate the national semiconductor industry in Taiwan (Shilov 2022). The rise of South Korea's semiconductor industry was also due to the same reason (Holbrook 1995). However, the barriers to entry in the semiconductor industry are becoming increasingly high even with government support, hindering the competitiveness of latecomers. China's semiconductor industry is an obvious example because too many new companies compete for limited government subsidies (Byun 1994). The situation is getting worse due to the US-China trade war (Tamirat 2017).
- **Science parks (industrial clusters)**: In addition to cultivating professional talents and technologies, science parks were also set up in Taiwan to further promote the development of the industry manufacturing industry. Under the influence of industrial clusters, Taiwan created a high-value semiconductor industry. Taiwan's semiconductor supply chains remain highly competitive despite the changing international situation. The South Korean government has taken similar measures. In January 2024, the South Korean government announced plans to invest approximately US$470 billion over the next 23 years to build a large-scale production park in Gyeonggi Province with SK Hynix to create the world's largest semiconductor cluster (Grimes and Du 2022).

During the US-China trade war, the United States sought to limit or delay China's ability to acquire advanced semiconductor manufacturing technologies and equipment (Fuller 2019). In the view of Shivakumar et al. (2024), new chip restrictions have severely impacted China's semiconductor ecosystem, limiting access to equipment critical to next-generation production. However, China has been forced to increase domestic investment in more advanced chips while also reducing the market share of U.S. companies, thereby lowering the revenue of U.S. companies as well as the funds needed to develop next-generation technologies. Over time, the loss of the market share is likely to weaken the competitiveness of U.S. companies in this critical industry.

The competitiveness of a country's semiconductor industry is directly reflected in chip exports. According to Rogers (2021), EU countries accounted for only 6.6% of global chip exports in 2019, lagging behind the 6.9% of the United States. Market leaders were South Korea and Taiwan, which held shares of 17.9% and 14.4%, respectively.

5.4 Sustainability and Influencing Factors

Sustainability is a method of consuming resources to meet current needs without compromising the ability to meet future needs (Wu et al. 2023). Sustainability has three dimensions: economic, environmental and social (Fig. 5.15). The sustainability of a company generally means that (Wu et al. 2023).

- The company continues to make profits or increase market share;
- The demand for the products or services provided by the company will not disappear. Even if it disappears, the company can replace them with other products or services in a timely manner.
- The company is able to withstand longer periods of losses than its competitors;
- The company consumes very little energy;
- The company does not cause any harm to the environment;
- The company does not deprive the rest of society (including those in the future) of available resources.

But this definition is based on the premise that the company must continue to operate for a long time.

According to TSMC, the world's largest foundry, to make semiconductor supply chains sustainable, companies need to conduct activities along the three dimensions of sustainability (TSMC 2024):

- Social:
  - Abide by ethics code;
  - Follow labor policies;
  - Have occupational safety and health procedures;
  - Manage fire and safety protection.

Fig. 5.15 Three dimensions of sustainability

- Environmental:
  - Reduce waste;
  - Save energy and water;
  - Reduce carbon emissions;
  - Prevent pollution;
  - Manage hazardous material usage.
- Economic:
  - Use a business continuity plan;
  - Enhance quality and processes;
  - Develop an energy saving tool;
  - Develop a circular economy;
  - Evaluate supply chain capacity and inventory;
  - Upgrade industry localization.

5.4.1 Sustainability Versus Competitiveness

Although a company's competitiveness can also be expressed as a long-term advantage over its competitors, competitiveness and sustainability usually imply short-term and long-term success for the company, respectively. Therefore, short-term competitiveness is considered to contribute to long-term sustainability. However, some recent cases have shown that it is possible to pursue long-term sustainability at the expense of short-term competitiveness, e.g., some recent cases of wafer foundries' supply chain localization plans.

Take TSMC as an example. While TSMC is building new localized wafer fabs in the United States and Japan, it also continues to build up more wafer fabrication capacity in Taiwan, since short-term competitiveness is still more important (or urgent) than long-term sustainability. Continuing with existing practices has proven to be the most efficient and economical, which is conducive to the short-term competitiveness. Nevertheless, companies with higher competitiveness are generally more likely to be sustainable (Wagner and Schaltegger 2003).

However, the competitiveness of a company does not guarantee its sustainability:

- In a market where a company is locked in, a reduction in demand for its goods will eliminate the sustainability of the company. For example, the long-term dominance of Japanese manufacturers in traditional cameras disappeared with the advent of digital cameras (Sendico 2024).
- From a consequentialist perspective, an increase in the market share means a higher competitiveness. Price-cutting competition is often used to increase the market share, but it cannot last long, that is, it is not sustainable.

On the contrary, usually a company that can sustain implies that it is quite competitive. However, a company's sustainability does not necessarily guarantee its competitiveness. For example, governments in some regions invest large amounts of budgets to support the development of certain industries (such as the semiconductor industry). These benefited companies are thus able to compete unfairly with other rivals (in other regions). However, after government support ended, these companies were no longer able to compete with their rivals. Semiconductor R&D and manufacturing companies in China are typical examples.

5.5 Competitiveness and Sustainability of Semiconductor Manufacturing Through Supply Chain Localization

5.5.1 TSMC Case

Singh et al. (2020) listed four sources for the competitiveness of TSMC. An analysis of whether these sources will be affected by TSMC's attempts to localize its semiconductor supply chain is as follows:

- Time and cost of building new wafer fabs: The United States and Japan have provided TSMC with substantial government subsidies, reducing TSMC's burden in this regard.
- Difficulties for new competitors to recruit sufficient skilled labor: TSMC may face the same difficulties after semiconductor supply chains are localized.
- Increasing high prices of advanced wafer fabrication equipment: Top chip designers often have close ties with advanced wafer fabrication equipment suppliers. So, by building new wafer fabs closer to TSMC's customers (top chip designers), TSMC may have easier access to such wafer fabrication equipment.
- Advanced packaging patents: Currently, the most advanced packaging companies are located in Taiwan and the United States, so building new wafer fabs in the United States in addition to Taiwan seems to be acceptable.

For TSMC, perhaps now is the best time to localize its semiconductor supply chain. Because with the craze of generative artificial intelligence (GenAI), the demand for the most advanced chips has exploded, and TSMC dominates the market supply in this field. Therefore, since the GenAI flow improves TSMC's competitiveness, it can tolerate a slight decrease in competitiveness caused by building new fabs in non-optimal regions, especially after securing subsidies and concessions from the central and local governments of the two countries.

5.5.2 Comparison Between TSMC and Samsung

The recent localization actions taken by two major wafer foundries, TSMC and Samsung, are used to illustrate how supply chain localization can help rebuild the competitiveness and even sustainability of semiconductor manufacturing.

First, for TSMC, whether it builds a new wafer fab in the United States or Japan, it can increase its fabrication capacity. When demand is greater than supply, more capacity helps win more market share. Therefore, assuming that competitors do not take the same action, TSMC is expected to further enhance its **competitiveness** in this way.

In addition, building new wafer fabs in the United States or Japan instead of Taiwan helps reduce the risk of war (i.e., geopolitics), which is obviously an effective measure to ensure the **sustainability** of TSMC. This also explains why TSMC was eager to build the second and third new wafer fabs after the completion of its first new wafer fab in Japan. However, TSMC continues to build new wafer fabs in Taiwan because the demand for fabrication capacity is still huge, and building new wafer fabs in Taiwan is the fastest way for TSMC to meet this demand, thereby ensuring **competitiveness**. Maintaining high **competitiveness** in the long term also contributes to **sustainability** (see Fig. 5.16).

A similar analysis can be done for Samsung, which pursues localization by building new wafer fabs in South Korea and also the United States to mitigate the impact of the US-China trade war (see Fig. 5.17). Owing to the US-China trade war, wafer fabs in China are prohibited from acquiring high-precision wafer fabrication equipment, which also affected Samsung's wafer fabs in China. In addition, domestic wafer fabs in China kept expanding their capacity, detrimental to the profitability of Samsung's wafer fabs in China.

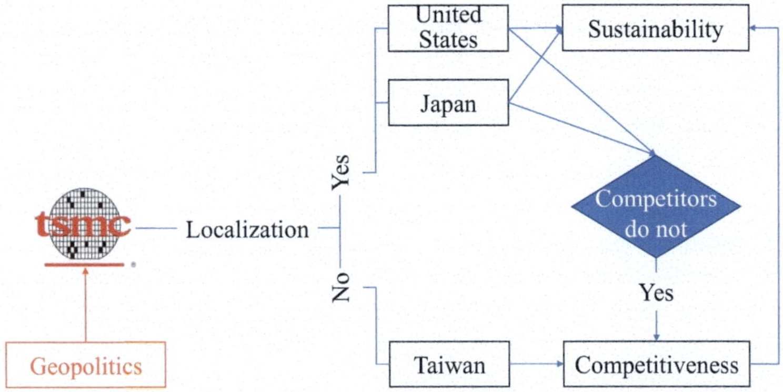

Fig. 5.16 Linking supply chain localization to competitiveness and sustainability for TSMC

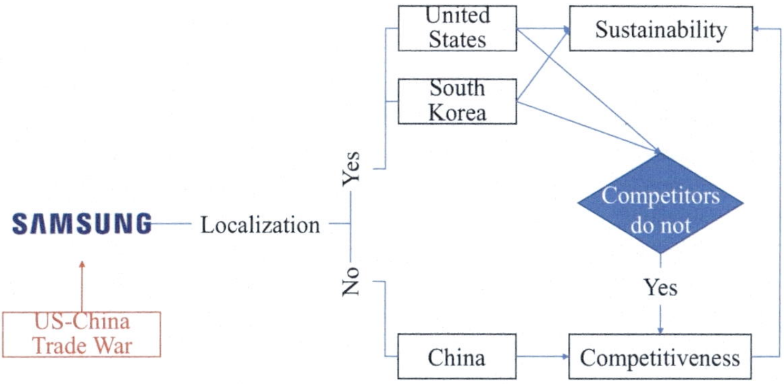

Fig. 5.17 Linking supply chain localization to competitiveness and sustainability for Samsung

Samsung listed four criteria that localized wafer fab locations must meet to ensure the competitiveness: talent recruitment, existing semiconductor manufacturing ecosystem, speed to market, and strong public–private partnerships (i.e. incentives) (James 2021). Samsung is considering building new wafer fabs in three states: Texas, Arizona and New York. Samsung has filed documents and negotiated tax incentives and breaks with authorities in Arizona, New York and Texas as it sought to build a new wafer fab that is expected to create 1,800 jobs and cost $17 billion.

Shilov (2023) compared the contributions of these three locations to Samsung's competitiveness as follows. Samsung already has a wafer fab in Texas, so the state already has an ecosystem of talent and suppliers. By comparison, it must compete for talent and supplies in Arizona (against Intel and TSMC) and New York (against GlobalFoundries). In addition, because Samsung already has a factory in Texas, it was unable to negotiate favorable tax incentives with Texas officials. Arizona was chosen for offering generous tax incentives and available land, which contributed to Samsung's competitiveness. Its final choice was Arizona due to its lower cost of living, lower personal and business taxes, more affordable housing, available land, and fewer regulatory hurdles, i.e., strong public–private partnerships (i.e. incentives).

Samsung has many customers in the US (such as IBM, Nvidia, Qualcomm, Tesla, etc.) that require access to its most advanced nodes and would prefer to use fabs located in the US, which is why Samsung is urging to build a wafer fab there. In addition, the company needs to build a new leading wafer fab in North America to better compete with its rival TSMC, which planned to build wafer fabs in Arizona by 2024. This plan may fail as it often takes more time to push more advanced fabrication processes to a profitable state, i.e. with sufficiently high yields. New fabs in the U.S. won't solve this problem, but existing fabs in South Korea are more likely to do so. This problem is no exception for TSMC.

Semiconductor manufacturers in other countries have taken similar actions to ensure their sustainability and even competitiveness. For example, in the third quarter of 2020, NXP of the Netherlands opened its newest wafer fab, a 150 mm gallium nitride (GaN) wafer fab in Chandler, Arizona, which the company confidently called "the most advanced fab in the United States for 5G RF power amplifiers" (Shepardson 2023).

The original intention of TSMC's and Samsung's localization attempts is to improve **sustainability**. TSMC is moving some of its wafer fabrication capacity away from Taiwan so that if a war breaks out across the Taiwan Strait, China will not be able to capture or destroy all of its capacity. In addition, the competition in China's semiconductor industry is becoming increasingly fierce. Samsung's wafer fabs in China may not be profitable in the future, so Samsung is building alternative wafer fabs domestically. In contrast, the tension on the Korean Peninsula is not the focus.

TSMC's Arizona fab plans to fabricate 4-nm and 3-nm chips, but they will not be packaged in the United States and will still have to be shipped back to Taiwan, seriously affecting the efficiency of the semiconductor supply chain. To solve this problem, following TSMC, semiconductor packaging and testing company Amkor also plans to establish the largest advanced packaging and testing factory in the United States in Arizona, USA, to serve customers such as TSMC and Apple nearby (Shepardson 2023). In addition, LCY, one of the world's largest semiconductor chemicals manufacturers, was also considering building a factory in Arizona, USA. The factory will be LCY's largest investment outside Taiwan, and if the plan is finalized and approved, it is expected to be completed by the end of 2023, which is well in line with TSMC's 2024 production goals (TrendForce 2024). In this way, the localization of the semiconductor supply chain will be more complete.

TSMC's Arizona fab plan has encountered difficulties such as union struggles, recruitment difficulties, subsidy delays, and occupational accidents. Semiconductor supply chain localization appears not to be a planning issue but a control issue, because the location of building a localized wafer fab has already been determined. Only how to obtain more subsidies and cooperation is the focus.

Also, this is not a question of decision-making, but of execution. The country (region) for the construction of new wafer fabs has been determined. There are also a limited number of cities that can be considered based on specific requirements. Therefore, the question is how to smoothly proceed and complete the construction of the new wafer fab.

As mentioned earlier, in countries (or regions) where foundries are trying to localize their semiconductor supply chains, possible competition from local semiconductor manufacturers poses a threat to the success of the localization plans. In fact, for these local semiconductor manufacturers, the emergence of localized wafer fabs from abroad is not conducive to their short-term competitiveness or even long-term sustainability. Therefore, some local semiconductor manufacturers may further sacrifice short-term competitiveness in pursuit of future competitiveness and long-term sustainability. To achieve this, they must be financially strong enough to sustain large short-term losses. A clear example is that chip manufacturer Intel announced in August 2024 that it would lay off more than 15,000 employees, a reduction ratio of more than 15%, to cut costs by US$10 billion. The company's annual research and development and marketing expenses have also been reduced by billions of dollars this year, and capital expenditures have also been reduced by more than 20%. The

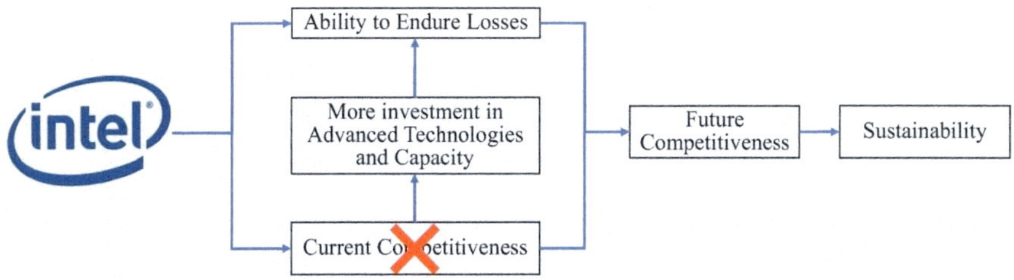

Fig. 5.18 Local semiconductor manufacturers' response measures in semiconductor supply chain localization: Intel as an example

purpose is to invest in new wafer fabs and extreme ultraviolet (EUV) exposure technologies to compete with TSMC, including its localized fabs in the United States (Nguyen 2024) (Fig. 5.18).

At first glance, the localization of semiconductor supply chains may appear uncompetitive due to higher salary levels, cultural differences, and other barriers. However, customers may demand to some extent that chips be manufactured in localized wafer fabs to avoid geopolitical risks. Although the emerging demand for high-tech artificial intelligence (AI) (Chen and Wang 2022a, 2024, 2022b, 2022c; Chen and Lee 2024b; Wang et al. 2023c; Wu et al. 2021; Yoo and Kang 2021; Zhang et al. 2021) chips requires readily available state-of-the-art wafer fabrication capabilities and has somewhat delayed localization attempts in the semiconductor industry, it still seems to be the way to go in the near future.

References

E. Armstrong, Principles for competitive semiconductor manufacturing. in *Proceedings of IEEE SEMI International Semiconductor Manufacturing Science Symposium'89* (1989). pp. 67–72

C.P. Bown, How the United States marched the semiconductor industry into its trade war with China. East Asian Econ. Rev. **24**(4), 349–388 (2020)

B.M. Byun, Growth and recent development of the Korean semiconductor industry. Asian Surv. **34**(8), 706–720 (1994)

T.C.T. Chen, Industry 4.0 for semiconductor manufacturing. in *Production Planning and Control in Semiconductor Manufacturing: Big Data Analytics and Industry 4.0 Applications* (2023), pp. 21–40

T.C.T. Chen, Big data analytics for semiconductor manufacturing. in *Production Planning and Control in Semiconductor Manufacturing: Big Data Analytics and Industry 4.0 Applications* (2023), pp. 1–19

T.C.T. Chen, Explainable artificial intelligence (XAI) in manufacturing. in *Explainable Artificial Intelligence (XAI) in Manufacturing: Methodology, Tools, and Applications* (2023), pp. 1–11

T.C.T. Chen, Applications of XAI for forecasting in the manufacturing domain. in *Explainable Artificial Intelligence (XAI) in Manufacturing: Methodology, Tools, and Applications* (2023), pp. 1–11

T.C.T. Chen, Applications of XAI for decision making in the manufacturing domain. in *Explainable Artificial Intelligence (XAI) in Manufacturing: Methodology, Tools, and Applications* (2023), pp. 51–81

T. C. T. Chen, Applications of XAI to job sequencing and scheduling in manufacturing. in *Explainable Artificial Intelligence (XAI) in Manufacturing: Methodology, Tools, and Applications* (2023), pp. 83–105

T.C.T. Chen, Job sequencing and scheduling. in *Production Planning and Control in Semiconductor Manufacturing: Big Data Analytics and Industry 4.0 Applications* (2023), pp. 77–100

T.C.T. Chen, Sustainable smart healthcare applications: lessons learned from the COVID-19 pandemic. in *Sustainable Smart Healthcare: Lessons Learned from the COVID-19 Pandemic* (2023), pp. 65–92

T.C.T. Chen, Cycle time prediction and output projection. in *Production Planning and Control in Semiconductor Manufacturing: Big Data Analytics and Industry 4.0 Applications* (2023), pp. 41–62

T.C.T. Chen, Defect pattern analysis, yield learning modeling, and yield prediction. in *Production Planning and Control in Semiconductor Manufacturing: Big Data Analytics and Industry 4.0 Applications* (2023), pp. 63–76

T.C.T. Chen, Y.C. Wang, Evaluating innovative future robotic applications in manufacturing using a fuzzy collaborative intelligence approach. The Int. J. Adv. Manuf. Technol. **130**(11), 6027–6041 (2024)

T. Chen, C.-W. Lin, Y.-C. Lin, A fuzzy collaborative forecasting approach based on XAI applications for cycle time range estimation. Appl. Soft Comput. **151**, 111122 (2024b)

T.C.T. Chen, Y.J. Lee, Smart technologies for fall detection and prevention. in *Smart and Healthy Walking: Toward Better Health and Life in Smart Cities* (2024), pp. 73–98

T.C.T. Chen, Y.J. Lee, Smart technologies for healthcare in smart cities. in *Smart and Healthy Walking: Toward Better Health and Life in Smart Cities* (2024), pp. 1–14

T.C.T. Chen, Y.C. Wang, AI applications to shop floor management in lean manufacturing. in *Artificial Intelligence and Lean Manufacturing* (2022), pp. 75–90

T.C.T. Chen, Y.C. Wang, AI applications to kaizen management. in *Artificial Intelligence and Lean Manufacturing* (2022), pp. 37–53

T.C.T. Chen, Y.C. Wang, Artificial intelligence in manufacturing. in *Artificial Intelligence and Lean Manufacturing* (2022), pp. 13–35

T.C.T. Chen, Y.C. Wang, C. W. Lin, Enhancing the effectiveness of output projection in wafer fabrication using an industry 4.0 and XAI approach. The Int. J. Adv. Manufact. Technol. **134**(1), 113–125 (2024)

T. Chen, Y.-C. Wang, Y.-C. Lin, Production planning and control of a pharmaceutical supply chain using evolutionary fuzzy approach. J. Eng. Res. 1–10 (2024)

S. Chiang, Chip giant TSMC opens first Japan factory as it diversifies away from Taiwan amid U.S.-China tensions. (2024). https://www.cnbc.com/2024/02/26/tsmc-opens-new-plant-in-japan-as-it-diversifies-away-from-taiwan.html

B.C. Crowder, Manufacturing science and manufacturing competitiveness. in *Proceedings IEEE SEMI International Semiconductor Manufacturing Science Symposium'89* (1989). pp. 32–33

A. Febransyah, J.I. Camelia Goni, Measuring the supply chain competitiveness of e-commerce industry in Indonesia. Competitiveness Rev.: An Int. Business J. **32**(2), 250–275 (2022)

D.B. Fuller, Growth, upgrading, and limited catch-up in China's semiconductor industry. in *Policy, Regulation, and Innovation in China's Electricity and Telecom Industries* (2019). pp. 262–303

O. Golovchenko, M. Saiensus, G. Sorokoumov, O. Onofriichuk, O. Zubko, L. Liu, Management of efficiency and competitiveness of enterprises. Econ. Aff. **67**(3), 317–326 (2022)

S. Grimes, D. Du, China's emerging role in the global semiconductor value chain. Telecommun. Policy **46**(2), 101959 (2022)

D. Holbrook, Government support of the semiconductor industry: diverse approaches and information flows. Bus. Econ. Hist. **24**(2), 133–165 (1995)

Invest Korea, Semiconductor (2023). https://www.investkorea.org/ik-en/cntnts/i-312/web.do

M. Ivanova, O. Varyanichenko, S. Sannikova, S. Faizova, Assessment of the competitiveness of enterprises. Econ. Annals-XXI **173**, 26–31 (2018)

L. James, Why are semiconductor fabs dotting Arizona, the "silicon desert? (2021). https://www.allaboutcircuits.com/news/semiconductor-fabs-dotting-arizona-silicon-desert/

T. Jenkins, F. Phail, S. Sackman, Semiconductor competitiveness in the 1990s. in *Proceedings Society of Automotive Engineers P223* (1990). pp. 249–255

D. Joshi, B. Nepal, A.P.S. Rathore, D. Sharma, On supply chain competitiveness of Indian automotive component manufacturing industry. Int. J. Prod. Econ. **143**(1), 151–161 (2013)

S.M. Khan, A. Mann, D. Peterson, The semiconductor supply chain: assessing national competitiveness. Center for Secur. Emerg. Technol. **8**(8), 1–98 (2021)

H.A. Lee, S. Korea to set up world's No. 1 semiconductor cluster in Seoul metropolitan area (2023). https://en.yna.co.kr/view/AEN20230315003751320

S.-H. Liao, T.-C. Hu, Knowledge transfer and competitive advantage on environmental uncertainty: an empirical study of the Taiwan semiconductor industry. Technovation **27**(6–7), 402–411 (2007)

G. Mcfee, The importance of competition in the semiconductor industry (2023). https://electronmagazine.com/the-importance-of-competition-in-the-semiconductor-industry/

J. Nguyen, Intel to cut 15% of its workforce (2024). https://www.abc10.com/article/news/local/intel-to-cut-workforce/103-9c66bb7a-728c-4166-8c6f-00a285521667

N.G. O'Connor, Z. Yang, L. Jiang, Challenges in gaining supply chain competitiveness: supplier response strategies and determinants. Ind. Mark. Manage. **72**, 138–151 (2018)

G. Papageorgiou, A. Efstathiades, T.G. Dimitriadis, Lean supply chain management and its impact on organisational competitiveness. Int. J Sup. Chain. Mgt **9**(4), 84 (2020)

C.-Y. Peng, C.-F. Chien, Data value development to enhance competitive advantage: a retrospective study of EDA systems for semiconductor fabrication. Internat. J. Services Tech. Managem. **4**(4–6), 365–383 (2003)

B.J. Jr. Pine, *Mass Customization: The New Frontier in Business Competition* (Harvard Business School Press, 1993)

M.E. Porter, How competitive forces shape strategy. in *Readings in Strategic Management* (1989), pp. 133–143

C. Rogers, Everyone loves localization of semiconductors. Applied Materials, Intel act. (2021). https://www.spglobal.com/marketintelligence/en/news-insights/latest-news-headlines/everyone-loves-localization-of-semiconductors-applied-materials-intel-act-63661421

S. Sarker, K. Rashidi, I. Gölgeci, D.M. Gligor, J. Hsuan, Exploring pillars of supply chain competitiveness: insights from leading global supply chains. Prod. Planning and Control **35**, 1025–1042 (2022)

Sendico, A comprehensive overview of the history of the Japanese camera industry (2024). https://sendico.com/blog/posts/a-comprehensive-overview-of-the-history-of-the-japanese-camera-industry

D. Shepardson, Amkor to build $2 billion Arizona semiconductor packaging plant. (2023). https://www.reuters.com/technology/amkor-build-2-billion-arizona-semiconductor-packaging-plant-2023-11-30/

A. Shilov, TSMC completes construction of 5nm fab 21 in Arizona. (2022). https://www.tomshardware.com/news/tsmc-fab-21-arizona

A. Shilov, Samsung foundry: New $17 billion fab in the USA by late 2023 (2021). https://www.anandtech.com/show/16483/samsung-in-the-usa-a-17-billion-usd-fab-by-late-2023

H. Shinno, H. Yoshioka, S. Marpaung, S. Hachiga, Quantitative SWOT analysis on global competitiveness of machine tool industry. J. Eng. des. **17**(03), 251–258 (2006)

S. Shivakumar, C. Wessner, T. Howell, Balancing the ledger: Export controls on U.S. chip technology to China. (2024). https://www.csis.org/analysis/balancing-ledger-export-controls-us-chip-technology-china

R. Singh, R. Garg, N.S. Patel, M.W. Braun, Generative adversarial networks for synthetic defect generation in assembly and test manufacturing. in *2020 31st Annual SEMI Advanced Semiconductor Manufacturing Conference* (2020), pp. 1–5

Y. Tamirat, The role of nanotechnology in semiconductor industry: review article. J. Mater. Sci. Nanotechnol **5**, 202 (2017)

TrendForce, [News] AMD eyes CoWoS-like supply for AI chips with TSMC full capacity (2024). https://www.trendforce.com/news/2024/01/03/news-amd-eyes-cowos-like-supply-for-ai-chips-with-tsmc-full-capacity/

M.L. Tseng, T.D. Bui, M.K. Lim, S. Lewi, A cause and effect model for digital sustainable supply chain competitiveness under uncertainties: enhancing digital platform. Sustainability **13**(18), 10150 (2021)

TSMC, A look at semiconductor supply chains (2024). https://www.tsmc.com/english/aboutTSMC/dc_infographics_supplychain

C.K. Udokporo, A. Anosike, M. Lim, S.P. Nadeem, J.A. Garza-Reyes, C.P. Ogbuka, Impact of Lean, Agile and Green (LAG) on business competitiveness: an empirical study of fast moving consumer goods businesses. Resour. Conserv. Recycl. **156**, 104714 (2020)

R. Vandendriessche, C. Buts, Closing the regulatory gap? A case study on the acquisition of a semiconductor producer under the EU foreign subsidies regulation. J. Europ. Competit. Law and Pract. **15**(2), 123–140 (2024)

A. Verma, N. Seth, Achieving supply chain competitiveness: some critical issues. Int. J. Sci. Eng. Technol. **2**(11), 6209–6213 (2010)

M. Wagner, S. Schaltegger, Introduction: How does sustainability performance relate to business competitiveness? Greener Manag. Int. **44**, 5–16 (2003)

S.T. Walsh, R.L. Boylan, C. McDermott, A. Paulson, The semiconductor silicon industry roadmap: epochs driven by the dynamics between disruptive technologies and core competencies. Technol. Forecasting Social Change **72**, 213–236 (2005)

Y.C. Wang, T. Chen, New XAI tools for selecting suitable 3D printing facilities in ubiquitous manufacturing. Complex and Intell. Syst. **9**(6), 6813–6829 (2023)

Y.C. Wang, T. Chen, M.C. Chiu, An explainable deep-learning approach for job cycle time prediction. Decision Anal. J. **6**, 100153 (2023a)

Y.C. Wang, T.C.T. Chen, M.C. Chiu, A systematic approach to enhance the explainability of artificial intelligence in healthcare with application to diagnosis of diabetes. Healthcare Anal. **3**, 100183 (2023b)

Y.C. Wang, T.C.T. Chen, H.C. Wu, A novel auto-weighting deep-learning fuzzy collaborative intelligence approach. Decision Anal. J. **6**, 100186 (2023c)

Y.C. Wang, T. Chen, Adapted techniques of explainable artificial intelligence for explaining genetic algorithms on the example of job scheduling. Expert Syst. Appl. **237**(A), 121369 (2024)

Wikipedia, List of semiconductor fabrication plants (2024). https://en.wikipedia.org/wiki/List_of_semiconductor_fabrication_plants

H.C. Wu, H.R. Tsai, T.C.T. Chen, K.W. Hsu, Energy-efficient production planning using a two-stage fuzzy approach. Mathematics **9**(10), 1101 (2021)

H.-C. Wu, T. Chen, M.-C. Chiu, Assessing the sustainability of smart healthcare applications using a multi-perspective fuzzy comprehensive evaluation approach. Digital Health **9**, 2055207623120390 (2023)

H.C. Wu, T.C.T. Chen, S.Y. Wang, Impact of the COVID-19 pandemic on the revenue of the catering industry: Taiwan as an example. SAGE Open **14**(2), 21582440241241410 (2024)

S. Yoo, N. Kang, Explainable artificial intelligence for manufacturing cost estimation and machining feature visualization. Expert Syst. Appl. **183**, 115430 (2021)

M. Zhang, F. Tao, A.Y.C. Nee, Digital twin enhanced dynamic job-shop scheduling. J. Manuf. Syst. **58**, 146–156 (2021)

The manufacturer's authorised representative in the EU is Springer Nature Customer Service Centre GmbH, Europaplatz 3, 69115 Heidelberg, Germany. If you have any concerns regarding our products, please contact ProductSafety@springernature.com

Printed and bound by CPI Group (UK) Ltd, Croydon, CR0 4YY

01/10/2026

02225980-0007